THE CRISIS OF LONDON

London is in a mess. This is evident from the increasingly unpleasant experience of daily life in the capital, from homelessness and unemployment to frustrating transport facilities and the general bad quality of the environment. However it is not only citizens of London who are suffering but the business community as well. London is having to face increasing competition from other European cities.

There is growing appreciation and debate about these problems from companies, political parties, local government and community organizations. *The Crisis of London* provides a solid analysis of what has gone wrong and explores policy directions that could make the city a more humane and livable place. Beginning with a discussion of the basic elements of a home, a job and a means of travelling around, it becomes clear that even in these essential aspects London is failing. A feature of the crisis is an increasingly divided city with conditions for the poorer citizens worsening all the time. The second section of the book concentrates on the quality of the environment. The authors explore the greening of the city and the need for sustainability, the privatization and dehumanization of public spaces, the fear experienced by women, denying them full access to the capital, the position of ethnic minorities, and the perspectives of local communities.

Employing the case studies of Docklands and King's Cross, the book raises the crucial question of the government of the capital. This review of the city concludes with an analysis of a potential vision for London involving both the creation of the necessary institutional structures and also the will to address the needs of all the capital's citizens. The book argues that a strategic approach is needed which accepts that the market alone cannot solve the problem. Stronger public intervention and government action is necessary if London is to match the developments in other European cities. Finally, such action has to be democratic and based upon full community involvement.

THE CRISIS OF LONDON

Edited by
Andy Thornley

London and New York

First published 1992
by Routledge
11 New Fetter Lane, London EC4P 4EE

Simultaneously published in the USA and Canada
by Routledge
a division of Routledge, Chapman and Hall, Inc.
29 West 35th Street, New York, NY 10001

Typeset in 10/12pt September by Leaper & Gard Ltd, Bristol
Printed and bound in Great Britain by
Biddles Ltd, Guildford and King's Lynn

British Library Cataloguing in Publication Data

A catalogue record for this title is available from the British Library.

Library of Congress Cataloging in Publication Data

The Crisis of London / edited by Andy Thornley.
p.   cm.
Includes bibliographical references and index.
ISBN 0–415–06885–1 (HB). – ISBN 0–415–06886–X (PB)
1. Urban policy–England–London.   2. Environmental policy–
England–London.   3. Community development, Urban–England–London.
4. Quality of life–England–London.   I. Thornley, Andy.
HT133.C72   1992   91–46095
307.76′09421–dc20   CIP

London, drawn by Duncan Thornley, aged 7

To Duncan and the next generation
of Londoners

CONTENTS

CONTENTS

FIGURES

TABLES

CONTRIBUTORS

All the contributors have written their chapters in a personal capacity and therefore their views do not necessarily reflect those of their employers.

Andy Thornley lectures at the University of Reading. His previous work as a professional planner included periods at both strategic and local levels in London. He is the author of *Urban Planning under Thatcherism: the Challenge of the Market*, published by Routledge.

Sue Brownill and **Cathy Sharp** worked on the Greater London Housing Study at University College London, which was funded by the Rowntree Foundation as part of their housing finance research programme. Sue now lectures at Oxford Polytechnic and is author of *Developing London's Docklands*, published by Paul Chapman. Cathy is now Research Manager at SHAC (The London Housing Aid Centre).

Andy Coupland lectures at the Polytechnic of Central London and is a freelance journalist, lecturer and tour guide, mainly on Docklands issues. He previously worked for five years as a community planner for the Joint Dockland Action Group.

Ruth Bashall and **Gavin Smith** are researchers at the Centre for Independent Transport Research in London (CILT). Gavin is the author of *Getting Around*, published by Pluto Press.

Duncan McLaren is a senior research officer at the Friends of the Earth Trust (26–28, Underwood St, London N1) and specializes in countryside, land-use and transport issues. His past work has included an environmental critique of government inner-city policy and collaboration with Tim Elkin and Mayer Hillman on *Reviving the City: Towards Sustainable Urban Development*, published by Friends of the Earth in association with the Policy Studies Institute.

John Punter lectures at the University of Reading. He has written extensively on urban design issues and published two in-depth studies of postwar design control for Reading and Bristol. He is author of *Design Control in Bristol: 1940–90*, published by Redcliffe Press.

Gill Valentine works for BBC Wales. She completed her Ph.D. at the University of Reading and spent some time as a journalist in London. Despite having left academia she continues to have an interest in feminist geography and writing about women's fear of crime.

Malcolm Cross is a Principal Research Fellow at the Centre for Research in Ethnic Relations at the University of Warwick. He previously worked at the Universities of Surrey and Aston and has over twenty years' experience of research into questions of race and ethnicity. He has written extensively on the topic, including more than a dozen books. He is editor of *New Community: A Journal of Research and Policy on Ethnic Relations*.

George Nicholson works for the Campaign for Homes in Central London (CHiCL) and is chair of the London Rivers Association. He was a councillor during the GLC's final administration and chaired its planning committee.

Michael Hebbert lectures at the London School of Economics. He edited (with Tony Travers) the *London Government Handbook*, published by Cassell, and contributed to *London: A New Metropolitan Geography*, edited by Hoggart and Green and published by Edward Arnold.

Michael Edwards lectures in the economics of planning at the Bartlett School, University College London. Since 1987 he has worked as an advisor to the King's Cross Railway Lands Group, KXT, and the London Boroughs of Camden and Islington.

Peter Newman lectures at the Polytechnic of Central London and is conducting research on urban politics in Britain and France. **Kelvin MacDonald** also lectures at the Polytechnic of Central London. They are both regular columnists for the journal *Planning*.

PREFACE

There are many books on London but most of these are oriented towards a popular, often tourist, audience and they do not provide much analysis of the processes of change that are taking place. On the other hand there are books which provide a very detailed description from one particular perspective, such as historical or geographical. The idea for a book on the 'Crisis of London' began in a discussion over puddings one lunchtime in the Polytechnic of Central London canteen, at the beginning of 1990. At that time a debate was starting to develop in a variety of different organizational settings, over the future of London, and this has mushroomed over the subsequent years. It was felt that there was a need for a book which would provide a solid basis of analysis and understanding which could inform these debates. The book should attempt to break free from the constraints of academic disciplines and explore the broad range of issues that affect the quality of life in the capital. It was also thought that the analysis should have a policy direction and help inform the debate over how London could be made into a more humane and livable city. The contributors to the book have been drawn both from academic backgrounds and from organizations that are developing ideas for London's future. It is expected that the book will be of interest not only to students of urban affairs but to anyone who is keen to know more about the processes of change that have been occurring in the city over the last decade and is concerned for the capital's future.

I would like to thank all the contributors who have done such a good job in meeting the brief asked of them and especially Peter Newman and Michael Edwards, who provided so many ideas beyond their particular sections. Finally I am very grateful to Sheena Gordon for her enthusiasm and help in pulling the final product together and Tristan Palmer, editor at Routledge, for his encouragement of the project.

Andy Thornley, London, July 1991

1

INTRODUCTION

Andy Thornley

London is in a mess. This view is reinforced for London's citizens in their everyday experiences. The problems of economic survival for the unemployed, underemployed or low-paid get more and more difficult as the costs of daily life increase. The number of homeless people sleeping on the streets is a highly visible sign of the crisis, as affordable housing becomes a rare commodity. Prices are increased to reduce the number of passengers using the underfunded and over-stretched transport system, although when usage falls again in the recession the response is the same and prices go up further. Thus the simple process of moving about the city becomes a luxury. People, usually women, making their trips to shops or schools during the off-peak hours have to endure long waits at bus stops with litter swirling about their feet. At night, women and the elderly don't even feel able to go out on their own. Those who have a job endure long, uncomfortable and erratic journeys on a transport system in which no-one has any confidence because of frequent disasters. Alternatively they might try driving and spend unproductive hours in traffic jams on roads full of pot-holes and subject to frequent diversions because of burst water mains. Life in London is stressful. London as a city is falling down the quality-of-life league tables and it is not surprising that a recent London Weekend Television survey showed that 48 per cent of Londoners would like to leave.

This is the everyday image of crisis. The aim of this book is to explore beneath this experience and find out in more detail what is happening, what exactly is going wrong and what might be done about it. Such an exploration has to be conducted against the preference in some quarters to try to ignore the problem. Central government, investors and developers want to play it down. Image and confidence is an important element of their business. Crisis, riots, homelessness on the streets: these are images that have to be suppressed and replaced by that of a vibrant, thriving metropolis – according to the British director of the developers Olympia & York, Canary Wharf presents a 'beacon of hope' for all Londoners. So are we citizens of London making too much of the problem and displaying a typical British tendency for moaning instead of enjoying London's attractions and quality?

The crisis in the quality of life in London has also been expressed as a crisis of

1

government. London ever since the demise of the GLC has had no single authority. There has been no voice expressing the concerns for London as a whole, extolling its image or dealing with its strategic problems. Although no-one is advocating a return to the large bureaucracy of the GLC there is widespread feeling, from developers to local communities, that there is a need for some kind of body to play this role – though, as we shall see, there are different ideas on how this should be done. However, the crisis of government goes beyond the city scale and extends both upwards and downwards. At the broader level the ideological importance given to market processes by governments over the last decade and more has created its own difficulties. The result has been a very 'light touch' given to regional and strategic planning. This has resulted in overheating in the South-East and no clear and positive strategy for future developments in the region. This reliance on market processes with their short-term horizons and inherent failure to co-ordinate or generate infrastructure could be said to have left London without the policy framework for investing in the future. At the local scale we have London boroughs, many tackling enormous social problems, struggling to prioritize which aspects of their services they have to cut. This financial straitjacket on public expenditure has an obvious effect on quality of life. In any case there is the problem of the boroughs being so large that they do not match people's feeling of community identity. Therefore there is also a crisis of democracy in London.

It has become a popular journalistic pastime to compare the crisis of London with the vitality of Paris. Paris is exciting, full of life, small shops throughout the centre, fresh baguettes every morning and dramatic architecture. It also has the ability to plan for the future. It believes in the importance of public intervention and has been expanding its transport network and co-ordinating it with development projects. It has connected the city to the rest of Europe through the high-speed rail network, the TGV, and linked it to the new and expandable airport, Disneyland projects, international hotels and conference centres. This epitomizes the role that the public sector is playing in trying to make Paris the Gateway to Europe and its cultural centre. Contrast this with the muddle over the Channel Tunnel links. However, once again it is necessary to explore below the surface. This middle-class, café-society perspective on Paris neglects the social impact that has been taking place. The projects have displaced communities out to the periphery, a process that is planned to continue with the displacement of the poor African community of the Secteur Seine Sud-Est. France is beginning to have serious problems with these peripheral estates – problems which of course are not visible to tourists and businessmen in the centre. Another loser is democracy. The French have a very centralized and technocratic approach to decision-making based upon their history and the élite educational system, the *grands écoles*. Thus the comparison is much more complex than might at first appear and raises issues of the relationship between public intervention, positive action, democratic processes and which citizens gain and lose. These are issues which will be explored throughout this book.

The discussion of Paris raises another major issue that London faces, namely the increased competition from European cities in the light of the Single Market (Coopers & Lybrand Deloitte 1991; Cheshire 1990). The prosperity of London has relied for a very long time on its financial rôle. This is now under threat and vigorous attempts are being made by other cities to capture this rôle. The strongest contender in the field is Frankfurt, as the location of the powerful Bundesbank. This position will be strengthened if, as is likely, the European Bank is located there. London as an office location is also vulnerable with its current prime office rents far in excess of any other European city. The new transport geography of Europe is crucial here. The high-speed network, and new airports such as those of Paris and Munich are changing the pattern of travel both to Europe and within it. London could lose much of its traffic as the main airport of entry to Europe, and whereas other cities are linked into the high-speed rail network London is out on a limb, and for a long while it will be down a rather slow branch-line. Competition will increasingly be felt from Paris, Brussels (benefiting also from the move of the European Parliament), Frankfurt, Munich and Milan. The closer links with former East European countries and the addition of Scandinavia to the EC will move the centre of gravity eastwards and, once its short-term problems are overcome, Berlin, the capital of a united Germany, will also provide a strong challenge. In this highly competitive environment the quality of life and image of each city will play a more and more important rôle. Good housing, pleasant environment, lively culture and efficient transport will all play their part. It can be seen that London has a lot of catching up to do. Thus overcoming the crisis is not only important in order to make life more bearable for Londoners: it is also essential for the economic survival of the city.

London will also be facing competition for investment from provincial cities within Britain, many of whom have been actively promoting their image. The extent of the crisis in London will have its ramifications for these cities. People, and companies, are attracted by what they see as a better quality of life in the smaller cities, with less congestion and easier access to open countryside. The continued failure of London to deal with its problems will increase the relative attraction of other cities. Peter Hall (1991) has expressed the view that provincial cities will experience a golden age from the mid-1990s. On the other hand it is sometimes said that in the increasing competitiveness of the international market Britain will have to 'sell' its most competitive product and that this points to increasing investment in London as the only city with sufficient potential at this world scale. A similar argument has been made in some quarters in Sweden for placing less emphasis on regional policy and investing in Stockholm.

Within London there is of course enormous variety. Although it is located in the prosperous region of the country, there is great poverty in the capital. One feature of this poverty is its geographical structure. In contrast to some European cities such as Paris or Rome much of this poverty is quite centrally located. This results in the close juxtaposition of some of the poorest and some of the wealthiest communities. Such a contrast is particularly evident in Docklands, where, alongside the symbols of affluence and dynamism such as Canary Wharf and post-modern luxury housing one gets some of the worst housing estates in Britain, falling apart owing to lack of maintenance. It is about ten years since the city experienced the violent uprisings in Brixton and Tottenham but the underlying conditions still remain. Racism is still rife. Unemployment in parts of London continues to be very high – in 1989 it was 14.6 per cent of the economically active population in Hackney, 13.5 per cent in Tower Hamlets and 11.7 per cent in Lambeth (London Research Centre 1989). The differences across the capital are striking – the comparable figures for the boroughs of Kingston, Hillingdon, Sutton, Richmond and Harrow were all under 3 per cent. As the London Research Centre has shown, the divergence between London's highest and lowest income groups continues to grow. In 1987 the average gross weekly income of the lowest 10 per cent of households was about £57, being an increase of 30 per cent since 1980. The equivalent for the top 10 per cent was about £670, having increased 118 per cent over the same period. What indication is there here of the 'trickle-down' effect of the market approach?

Such contrasts also have a geographical dimension in the ghettoization of the city. The poorer citizens have lost their rights over parts of the city – more and more areas are excluded as they cater only for those with higher income and tourists – and they are faced also with the disincentive of extremely high public transport costs. These areas are becoming increasingly privatized, with gated enclosures and policed indoor courtyards. The public realm shrinks (Punter 1990) and a fortress mentality based upon fear develops. Such tendencies could increase, with a severe effect on London's image. The attraction of many European cities is their cosmopolitan nature, which contributes so much to the culture, variety and vitality of their activities and use of public spaces. This could be London's future, but it depends upon what action is taken. A continuation of the market-led philosophy will only accentuate the social disparities across London. In which case a better comparison to make would be with New York (see, for example, Savitch 1988; Fainstein 1990; New Statesman and Society 1991). The indications are that we could look forward to more extreme cuts in services, increased violence and social problems, and further deterioration of infrastructure. Parts of London, such as Hackney or Lambeth, could become areas where no-one cares and which are abandoned by central government and the private sector. This scenario can be illustrated by the issue of safety. It will be seen in Chapter 7 of this book that one of the urgent needs in London is to improve the safety of the city's streets for women. In New York cuts are being made in the lighting budget and one in four street lights are to be turned off. We enter the downward spiral of deteriorating environment and crime.

The reliance on market processes has other consequences. We have witnessed the way in which organizations from health authorities, water boards and British Rail to local authorities have been compelled to sell off land and property assets in order to maintain their levels of public service. Planning gain has also become the prime means by which community benefits can be achieved. As we shall see later in the book the property market is very fragile and volatile and full of uncertainty. Linking the provision of basic services and community needs to this market can therefore have extreme consequences.

However, the message of this book is that there is nothing inevitable about London's future. The pressures of world economic changes cannot be ignored but they can be moulded and shaped in different ways (Gurr and King 1987). A shift of ideological perspective and political will can save us from New York's fate. London is at a critical juncture and the direction of public policy over the coming years will be crucial.

So it can be seen that the crisis of London has many dimensions. The quality of the environment in the city is very low, the streets are dirty, while pleasant, accessible public spaces are difficult to find. However, for the citizens of London the crisis encroaches on the very basics of life: affordable housing, a job and decent income, and the ability to move about. The structures of decision-making and resourcing are just not up to the task. There is a crisis of government and democracy. However, these problems, directly and sharply felt by those who live in London, should also be of major concern to companies and business. Image and confidence has to be based upon reality. London can no longer afford to drift down on the market tide, still drawing its energies from its long-lost imperial past. There is a need to appreciate the severity of the crisis and mount a positive and co-ordinated attack to address it. The approach being taken by London's competitors in Europe should provide the stimulus.

In the following chapters this book will set out the nature of the crisis in each aspect of the city's life. Of course it is artificial to chop up the crisis into these separate elements, but this is unavoidable if we want to gain a detailed understanding. The interconnected nature of the crisis will be addressed in the final chapter, together with some pointers to the future. The aim of the book is to provide a detailed analysis of the nature of London's problems as a basis for discussing new policy directions that can ensure London's future as a city worth living in. It therefore differs from other recent books on London. Hoggart and Green (1991) have provided us with a very useful collection of material on London. This provides information on past trends in the city with a focus on economic and spatial changes and is a valuable foundation on which to build the discussions that are of central concern to us here. The difference in emphasis between the two books is illustrated by the contributors. In Hoggart and Green all the chapters are written by academic geographers, whereas the policy orientation of this book is reflected in the decision to draw upon the knowledge and experience of people who are working to influence London's future. Peter Hall's book *London 2001* has more of a policy orientation. However, his remit is the whole of the South-East and he is therefore unable to treat some of London's problems in detail. His perspective is also a rather personal one, although admittedly one with considerable influence! He claims that his book is apolitical, and it could be said that he is able to make this claim because of his selection of

5

material. His discussions are mainly in terms of spatial strategies, with less concern for social and distributional issues. In the following chapters we will be taking a broad approach to London's problems including these issues and others not included in the above literature, such as safety, ecology and urban design – important issues affecting the quality of life for Londoners.

The first part of the book examines the very basics of life in London: a home, a job and a means of travelling around. It is clear that here we have some of the most extreme aspects of the crisis; yet the satisfaction of these needs should be seen as a fundamental right of citizens. In their chapter, Sue Brownill and Cathy Sharp demonstrate the severe problems in London's housing provision. This results, amongst other things, in homelessness, lack of affordable housing and a divided city. They trace the impact of national housing policy, demonstrate the failure of a reliance on a market approach and set out principles for an alternative. They also make the point that the crisis in London's housing has severe effects on the viability of London's economy. This is the topic of Chapter 3, in which Andy Coupland outlines the dramatic shift in London's economic structure in recent years leading to the almost total collapse of the manufacturing sector. He describes the impact of the 'big bang' and analyses the operations of the speculative office market. The result of these trends is a highly volatile and unstable economy with great spatial variations across the capital. Another effect of the trend is an increase in commuting distances. Transport is the topic of Chapter 4. It would be difficult to find anyone who denies that there is a transport crisis in London; the constant jams on the roads and the costs and unreliability of public transport make this all too obvious. At the time of writing the government has just announced its Citizen's Charter, which includes some transport measures, such as privatization of British Rail, deregulation of London's buses and charging repair contractors for placing cones on motorways. In their analysis of the problems, Ruth Bashall and Gavin Smith show that such market-led solutions are just not up to the task. They demonstrate that under the symptoms of crisis which we all suffer, lie more fundamental problems of investment, organization, marketing and equity. They conclude with an agenda for action which would help to bring London's transport system up to European standards and provide a real citizen's charter: the right for all London's population and visitors to move about in comfort and safety.

The next section of the book focuses on the quality of the physical environment. The greening of the city has become a popular demand in recent years and there have been a lot of complaints about the loss of open space, litter and the effects of traffic pollution. In Chapter 5, Duncan McLaren argues that these issues must be viewed from the broader perspectives of urban ecosystem and sustainable development. The two concepts need to be integrated and London's environmental crisis placed in a world and national context. There is a need to escape from the imperatives of international capital investment and start from the perspective of a better environment for all London's inhabitants. A planning strategy is needed which is less wasteful of resources in building and transport. In

Chapter 6, John Punter switches our attention to the quality of the built environment and urban design. This has also attracted a lot of recent interest, particularly through the comments of Prince Charles. Much of this debate has been about the question of architectural style. However, John suggests that this has to be analysed in a wider context and he explores the nature of the property boom over the last decade and the new forms of development that have resulted. One of the main problems is the trend towards the privatization of the public realm. If this is to be tackled in the future a comprehensive approach to the planning of public spaces has to be adopted which includes political will, planning gain, design guidance and participation. A key component in such participation would be the involvement of women. The quality of public spaces and transport and the way this interacts with crime is of particular importance for women. London is not a safe place for women; a city which generates fear in the majority of its citizens cannot be regarded as civilized. These are the issues addressed by Gill Valentine in the next chapter. She shows how this fear is generated and how it limits women's freedom as they adjust their behaviour to cope. She also discusses the initiatives that are needed to overcome the problem of safety and which would allow women to enjoy their right of access to London.

The issue of race is intricately woven into the crisis of London. Malcolm Cross explores this issue in Chapter 8. He shows how the widening gulf between rich and poor, in both spatial and social senses, has affected different ethnic groups. He describes the spatial distributions of these groups and their representation in the changing labour market. One message that arises is the need to disaggregate the analysis by group, age and sex. He then explores various theoretical strands in order to help assess the importance of race and ethnicity for an understanding of the way London is changing. Ethnic groups are particularly vulnerable to current economic changes and some groups have additional disadvantages because of living in inner London. George Nicholson in the following chapter focuses upon communities in one part of London – those inner areas subjected to pressure from the expansion of office development. He outlines the history behind the GLC's Community Areas Policy and uses this to argue for the importance of the neighbourhood principle. He exposes the dangers of a top-down approach, whether in the form of unitary authorities, imposed structures or planning principles. Instead he puts forward alternative approaches which stress a diversity and range of ideas which draw upon the energy and participation of London's citizens. Michael Hebbert's task in Chapter 10 is to investigate to what extent there is a crisis of government in London. The first point he makes is that many of London's problems stem from broader trends and national policies and are therefore not related to the particular political arrangement in the city itself. He assesses the strengths and weaknesses of arrangements in London since the abolition of the GLC. He concludes that, although there has been an increase in the political polarization of the boroughs, there has also been an increase in civic pride. Notwithstanding their differences, good co-operation has occurred between boroughs, but this has been limited in its scope. It is at the level of

London as a whole that the crisis of government occurs. He reviews the various proposals to deal with this and concludes with a suggestion that the City Corporation could take on a new city-wide function.

The rest of the book seeks to examine the various dimensions of the crisis in a more interrelated way. First through the two case studies of Docklands and King's Cross. Docklands can be considered the flagship of the Thatcherite approach to dealing with London's problems, and in his chapter Andy Coupland explores whether it has succeeded in this rôle. It is an experiment in the market-led approach in which the public sector is expected to prime the pump, assemble land, provide infrastructure and market the area. Andy gives the history of planning in the area, outlines the approach of the London Dockland Corporation and the operation of the Enterprise Zone. He then gives a detailed account of Canary Wharf, which is a crucial element in the story. The conclusions that can be drawn reinforce the messages from previous chapters: a divided community, lack of affordable housing, inadequate transport infrastructure, a vulnerable local economy and the failure of the trickle-down effect. As overprovision of offices looms, the whole experiment does not seem to have anything to offer either to the local community or London as a whole. All this, and the costs of pump-priming continue to mount. Of course in King's Cross no development has yet begun. However, Michael Edwards is able to use the controversy over the proposals to illuminate in detail the forces that underlie development in London. He analyses the motives and actions of the various actors in the process: investors, landowners including British Rail, developers, local authorities, central government, professionals and community groups. He presents a very instructive account of the way the objectives and imperatives of particular interests can lead to decisions that affect the lives of all Londoners. Finally, in the last chapter, a further attempt is made to draw the material together. The focus this time is on ideas to help overcome the crisis. How can all the various needs and interests be brought together into a vision which can act as a vehicle for co-ordination and planning? What structures are needed for the formulation and implementation of the vision? How can local democracy and community involvement be fostered in the face of increased economic competition between cities? Challenging questions – the important thing is to have a full and open debate. We set out some aspects that we feel should be put on the agenda of that debate.

REFERENCES

Cheshire, P. (1990) 'The outlook for development in London', *Land Development Studies* 7: 41–54.

Coopers & Lybrand Deloitte (1991) *London World City*, London, Coopers & Lybrand Deloitte. A study commissioned by the London Planning Advisory Committee and co-sponsored by Greater London Arts and the London Dockland Development Corporation.

Fainstein, S. S. (1990) 'Economics, politics and development policy: the convergence of New York and London', *International Journal of Urban and Regional Research* 14 (4).

Gurr, T. R. and King, D. S. (1987) *The State and the City*, London, Macmillan.

Hall, P. (1989) *London 2001*, London, Unwin Hyman.

—— (1991) 'A new strategy for the South East', *The Planner* 77 (10).

Hoggart, K. and Green, D. R. (eds) (1991) *London: A New Metrpolitan Geography*, London, Edward Arnold.

London Research Centre (1989) *London in Need*, London, LRC.

New Statesman and Society (1991) 'London and New York', a special series of articles in the issues of 25 January and 1 February.

Punter, J. (1990) 'The privatisation of the public realm', *Planning Practice and Research* 5 (3).

Savitch, H. V. (1988) *Post Industrial Cities: Politics and Planning in New York, Paris, and London*, Princeton, Princeton University Press.

2

LONDON'S HOUSING CRISIS

Sue Brownill and Cathy Sharp

No Londoner or visitor to the capital can fail to be aware of people sleeping on the streets and the 'homeless and hungry' signs of the tube station beggars. These people are the most visible tip of the iceberg of London's homeless.

Yet homelessness is far from being the only housing problem facing Londoners. High house prices affect even those on above-average incomes. Access to rented accommodation is restricted by both supply and costs. Insecurity and the threat of homelessness arise in all sectors. Londoners often have to pay high prices for poor-quality and overcrowded accommodation. Certain groups have been particularly affected by these trends, especially those who are economically marginalized, including young single people, ethnic minorities, women and the elderly.

The result is a housing crisis that not only affects all Londoners but also permeates many other aspects of the condition of life in London and threatens its economic viability. In this chapter we shall examine some of the dimensions of this crisis, analyse its causes and put forward possible ways to solve it.

A HOMELESS CITY

The true extent of homelessness in London is not known. Those who fall outside a narrow definition of homeless based on their statutory right to assistance under Part III of the Housing Act 1985 are not included in the official statistics. Many homeless people will not approach a local authority, knowing that they have no rights to rehousing. These people are simply not counted. There are an estimated 3,000 'street homeless' in London alone. These are the most visible of the homeless population, which also includes those in temporary accommodation such as hostels, bed-and-breakfast hotels, refuges, short-life properties and accommodation leased by local authorities from private landlords.

In 1989, a total of 33,170 households were accepted as homeless by local authorities in Greater London. This represented a rate of acceptance per thousand of the population of 12.4, a higher rate than anywhere else in Britain. Acceptances in London between 1983 and 1989 had increased by 38 per cent.

In September 1990 official DoE figures showed that there were 28,040 house-

holds placed in temporary accommodation by local authorities, pending enquir-
ies into their circumstances or awaiting permanent rehousing. These figures
represent a 29 per cent increase over the previous year. The poor conditions in
bed-and-breakfast hotels have been well documented (Conway and Kemp 1985).
Facilities are basic, space for young children, cooking and washing is limited and
there is a lack of privacy. Hotels may be in another borough, causing problems in
maintaining normal education and child-care arrangements, health care, work
and local community support.

The cost of providing bed-and-breakfast accommodation in London in 1989
was an estimated £113m. (Association of London Authorities 1990a). In
response to these high costs, many London boroughs have developed alternative
forms of temporary accommodation, notably by leasing accommodation from
private landlords. Around 13,000 properties had been provided in this way by
March 1991. The future of these schemes is now uncertain, since new government restrictions on leasing activities have come into force, heralding a return to
greater use of bed and breakfast.

Whatever form of temporary accommodation is used, the increasing scale of
the homelessness problem means that in practice temporary accommodation
often becomes 'home' for months or years. Most people in temporary accommodation are those whom the local authority has a duty to rehouse. Principally
this means people with children or others in 'priority need' through vulnerability
or for other special reasons. In 1990, there were around 8,000 children living in
bed-and-breakfast hotels in London.

Local authorities do not have a duty to rehouse most single people, childless
couples and those deemed to be 'intentionally' homeless. Such people are the
'hidden homeless' and the true extent of this problem is not known. Estimates
put the numbers of single homeless people in London at between 64,500 and
78,000; these are sleeping rough, squatting, living in bed-and-breakfast or other
unsuitable temporary accommodation (Single Homeless in London 1989). In
addition, the London Research Centre (1990) estimates that there are another
250,000 people living unwillingly as part of someone else's household. Many of
this group are young people, women and people from ethnic minorities reliant on
benefits or low wages. Most do not qualify for council housing, and housing
association accommodation is scarce. Their only real prospect of an independent
home is in the declining private rented sector. High rents and large deposits make
this an expensive and insecure option. Single people in this position are likely to
lead a nomadic existence, reliant on the goodwill of friends to put them up for
short periods, with no privacy or independence and with detrimental effects on
their health, relationships, work and ability to study (Threshold 1990).

PAYING FOR HOUSING IN LONDON

While homelessness remains the most extreme indicator of London's housing
problems the impact of the cost of housing in the capital has been steadily

attracting attention. Massive house-price inflation, rising interest rates and higher rents across all tenures has meant not only that many are priced out of adequate accommodation but also that some households are finding it increasingly difficult to continue meeting their housing costs.

Affordability as an issue has therefore become central to discussions on London's housing. The reasons for this become clear when we look at what Londoners have to pay for their housing (Table 2.1). Average house prices almost doubled between 1985 and 1989 from £46,600 to £91,783. They fell back to £87,266 in 1990 but despite this owner occupation is still beyond the reach of many working Londoners. To obtain a mortgage for a one-bedroom flat at the 'rock-bottom' of the market during the second quarter of 1990 required a gross income of at least £314 per week (London Research Centre 1990). On this basis only about half of all men and a fifth of women working full time could afford to buy on their own. A fifth of couples working full time would not have been able to buy even the cheapest one-bedroom flat. Over 40 per cent of couples could not buy the cheapest two-bedroom flat at the time.

Average council rents rose by 41 per cent between 1985 and 1989. In April 1990, weekly council rents increased by an average of £4.32 (18 per cent), following the introduction of a new financial regime by central government. Similarly, housing association rents increased by 37 per cent between 1985 and 1989, way above the level of inflation, again as a result of new legislation. Such large increases have led to intense debate about what constitutes an 'affordable rent' (Sharp *et al.* 1990). Providing accommodation at rent levels which are affordable by those not receiving housing benefit, principally the low-paid, is becoming increasingly difficult for associations and councils. Such dilemmas raise the possibility that social housing, that is low-cost rented housing provided

Table 2.1 Increases in house prices, local authority and housing association rents

Year	Average house price[1]	Index	Council rent[2] £p.w. net unrebated	Index	Housing assoc. rent regs. £p.w.	Index
1985	46,600	100	17.45	100	20.77	100
1986	54,286	116	18.14	104	22.25	107
1987	69,579	149	18.89	108	24.61	118
1988	79,417	170	21.65	124	26.37	127
1989	91,783	197	24.62	141	28.64[4]	137
1990	87,266	187	28.96[3]	166	NA	

Source: Nationwide Anglia Building Society, *London Housing Statistics*, London Research Centre, *Housing and Construction Statistics* (DoE)
1. Prices for the first quarter of each year
2. At April
3. Estimate
4. First half only

by local authorities and housing associations, will become 'residualized' and house only those receiving benefit.

The deregulation of rents in the private rented sector following the 1988 Housing Act also means higher rents for tenants, restricting access and promoting greater reliance on housing benefit. Over the 1980s, fair rents (those covered by the Rent Act) have risen by more than average earnings and by substantially more than retail prices. Existing tenants have experienced increases of between 17 and 21 per cent on re-registration, bringing the average up to £46.42 per week in late 1989. Post-1988 Act tenants face higher rents at an average of £61.38 according to the Rent Officer. But these figures are likely to be underestimates. Other sources reveal rents ranging from an average of £73.85 per week for a room or bedsit to £195.23 for a large house (London Research Centre 1990).

The impact of these increases in housing costs is felt by all Londoners, but some are proportionately more affected than others. Rented housing in all sectors will only be accessible to those in housing need if housing benefit is available to meet higher rents. However, since changes in 1988 there has been a fall in the numbers receiving benefit. In 1988/9 only one half of council tenants in London received housing benefit (Brownill et al. 1990).

A recent survey showed that 50 per cent of households with a mortgage were spending between one-third and one-half of their incomes on housing costs (including rates and heating). At the same time 50 per cent of social tenants were paying between 20 and 41 per cent of much lower incomes (Sharp et al. 1990). More significantly some households had increasingly less money left after meeting their housing costs. Tenants, pensioners, women and non-white households in particular had lower residual incomes.

The same research found that nearly one-third of council tenants and 44 per cent of housing association tenants reported problems paying their housing costs, a higher rate than the other sectors of the London housing market, despite lower costs (Maclennan et al. 1990). Mortgage arrears and repossessions have risen dramatically since 1989 and consequently the numbers being accepted as homeless through repossession is also on the increase.

HOUSING AND THE LONDON ECONOMY

The impact on the London economy of these trends has increasingly become a cause for concern. This was a problem recognized by employers and industrialists long before government perceived it. Recruitment or relocation of workers was becoming difficult, wage demands were including a housing element and companies were having to pay more to subsidize housing costs. In 1988 the chair of the London Chamber of Commerce wrote to the Chancellor of the Exchequer stating: 'The high cost of executive housing in London and parts of the South-East deters even highly paid staff from moving south. At the other end of the scale, inexpensive housing for rent does not exist. Urgent action is needed'.

13

As workers moved further out to find housing they could afford the London commuting area grew to include areas as far away as Doncaster. Longer journeys on an overloaded and increasingly unreliable public transport system or an equally overloaded road system meant not only lost hours to employers but strains on workers and the environment.

Longer journeys are not possible, however, for those who cannot afford the fares – particularly for women who want to work near home or part time. The provision of low-cost housing in the city centre for the many low-paid service workers such as those in hotel, catering, retail and tourist employment is vital for the functioning of the economy. London has traditionally had a large amount of council housing within and near the central area. As the supply of this housing has decreased and as central-area building has increasingly concentrated on the luxury end of the market the provision of low-cost central city housing has become an issue.

The problems for other sectors of workers, especially nurses and teachers, who are essential to the provision of high-quality services and facilities in the city, have been highlighted (Campaign for Homes in Central London 1988). For example nurses earning £170 per week cannot afford the £314 per week needed to buy. At the same time they are not in high enough priority need for social rented housing or housing benefit. As a result they may well have to spend as much as 43 per cent of their salary on privately rented accommodation (Crane 1990). The level of concern at this situation extended from the individuals involved to people living in inner urban areas who wanted a decent health and education service, to housebuilders who knew that the provision of good services was important for selling houses and to employers who wanted to attract and keep a workforce.

The foundation of London's severe housing problems can be found in a number of interacting factors. Obviously what has happened in London was a result of forces operating on a national and international level including, most importantly, economic change and restructuring, national housing policy and socio-economic divisions in the population. But these were added to by specific London factors and characteristics.

NATIONAL HOUSING POLICY AND ITS IMPACT ON LONDON

The rôle of housing policy in underwriting the trends and issues described in the previous section is shown clearly by Table 2.2. The table shows clearly that the main impact of existing policies on the supply of housing in London has been to fuel tenure shifts rather than promoting new supply. Overall, between 1981 and 1988 the stock increased by only 2 per cent, but the changes between tenures were dramatic. Council housing declined by 13 per cent and owner occupation increased by 23 per cent over this period. The long-term decline of the private rented sector, which in 1965 housed over half London's population, is also confirmed. The main reasons for these shifts were the relative financial advant-

Table 2.2 Trends in dwelling stock (000s)

Tenure	1981	1988	% change
Local authority	857	742	−13
Owner occupied	1343	1651	23
Housing association	117	149	27
Private rented & other	402	240	−40
Total	2719	2782	2

Source: *Housing and Construction Statistics* (DoE)

ages of each tenure promoted by finance policy. Within these changes it is significant that it is the supply of low-cost housing which has been declining.

Throughout the 1980s London local authorities saw a massive reduction in their ability to finance capital investment. Permission to borrow through the annual Housing Investment Programme (HIP) allocations made by central government was severely reduced. The ten years from 1979 saw a reduction of 84 per cent in real terms on HIP allocations to London boroughs to just £291m. in 1989/90. As a result the completion of local authority dwellings fell from over 15,000 to under 1,000 per year over the same period. The right-to-buy legislation, underwritten by large discounts and mortgage subsidy, had a drastic effect on council stocks. Between 1980 and 1989 more than 150,000 dwellings were sold in Greater London, 17 per cent of the April 1980 stock. Despite this high level of sales, restrictions on the use of capital receipts meant only small proportions of the revenue raised could be reinvested in housing.

The government's championing of housing associations to provide sufficient numbers of social rented housing has been shown to be over-optimistic. While numbers have increased they in no way make up for the reduction in council stock.

In a detailed assessment of housing need and supply, the London Research Centre estimates that in 1990 there was a shortfall of 380,500 social rented dwellings in London. This was estimated to rise by a further 264,500 over the five years to 1995 to give a total of 645,000. It is these shortages that are behind the increasing numbers of homeless and of households in temporary accommodation. The declining number of vacancies that have arisen in the shrinking stock have overwhelmingly been allocated to the homeless. In 1989/90, 61 per cent of all new lettings went to the homeless. Households on the general waiting list have very little prospect of a local authority dwelling and it has been estimated in some boroughs that it would take over 20 years to house just those on the waiting list at present, let alone new applicants.

Central government subsidies to finance public sector investment have also been drastically reduced. General subsidies to London local authorities fell by 44 per cent in real terms between 1981/2 and 1987/8. The result has firstly been higher rents, although until recently 'creative accounting' techniques to some extent softened the blow. Secondly, there has been a shift in subsidies from bricks and mortar to individual housing benefit, an outcome intended by government. Housing benefit subsidies increased in real terms by 325 per cent over the period 1981/2 to 1987/8. But as other subsidies fell, total subsidies to local authorities fell by 4 per cent in real terms over the same period.

London boroughs have been hit particularly hard by the 1989 Local Government and Housing Act. The resulting new financial regime has eradicated 'creative accounting', linked rents to house prices and attempted to make council housing entirely self-financing. By ensuring upward pressure on rents this system will continue the established trends of the 1980s and will end all other options for a local authority to pursue a housing strategy at odds with central government policy.

While local authorities have seen a decrease in government support the ideological fervour for owner occupation has been financially supported by increasing government subsidies to individual purchasers. The total bill for mortgage tax relief (MTR) grows as the numbers of owner occupiers increase and house prices and interest rates rise. Between 1981/2 and 1987/8 MTR increased by 136 per cent in real terms from £684m. while local authority subsidy fell by 44 per cent. This represented twice the amount of general subsidy going to local authorities. In 1988/9 MTR was estimated to have reached £987m. as interest rates rose. Yet despite the injection of massive sums of money to subsidize its purchase the result has not been a fall in prices nor have we seen a substantial rise in supply of new owner-occupied dwellings.

Private housebuilding completions have failed to make up for the gap left by the public sector slowdown in building. Despite the fact that private completions rose from 4,426 in 1980 to a high of 11,312 in 1988, total completions in the capital fell from 23,000 to 12,500 over the same period. The expansion of the owner-occupied stock has come through tenure transfers rather than new building.

The policy of directing subsidies at the individual purchaser whilst the market is trusted to provide the right quantity of housing at a cost affordable by the majority has been shown to be woefully lacking in the 1980s. Not only has the market failed to provide low-cost housing but in the late 1980s increased interest rates and stagnating house prices have dissuaded many existing owner occupiers from putting their homes on the market and has reduced new building starts.

Housing policy towards the other main tenures – the private rented sector and housing associations – has encapsulated the theme of promoting market relations in housing. The 1988 Housing Act sought to increase the role of the private sector by deregulating rents and introducing new forms of private investment through the Business Expansion Scheme (BES). Initial results of the BES have not been encouraging. Under 1,000 units per year have come on stream at rents of £400–£800 per month. The same Act laid down that housing associations had to include a certain proportion of private finance in schemes, the loan repayments leading to an upward pressure on rents.

The lack of investment in new building encouraged by the housing finance system has been matched by a lack of investment in improving the quality of the housing stock. Despite the fact that councils have been able to spend the capital receipts from the sale of dwellings on repairs, levels of investment have not been maintained in the face of HIP cuts and an ageing stock. In the owner-occupied sector there is evidence that owners who spend a high proportion of their income on their mortgage and those heavily affected by interest rate increases have been unable to maintain their properties.

In summary the impact of national housing policy has been to discourage investment in new dwellings and maintaining house conditions and to increase the costs across all tenures. Policies have also directly contributed to inequalities in housing provision and access to affordable housing. Not only is there a bias between tenures, but more importantly within the population. The structure of subsidies ensures that those on higher incomes receive higher than average subsidies. Those already advantaged in the housing market gain further financial advantages. Thus in 1988/9 higher-rate taxpayers received on average £1,025 per year MTR while standard-rate taxpayers received £480 per year (Brownill *et al.* 1990). Lower-income groups excluded from home ownership receive only means-tested housing benefit. Figure 2.1 shows the combined effect of untargeted mortgage tax relief and means-tested housing benefit. The dip in the middle

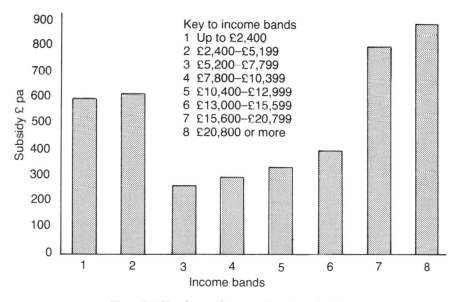

Figure 2.1 Total annual average housing subsidies

Source: Brownill *et al.* 1990.

represents those who are ineligible for housing benefit but who cannot afford to buy.

Subsidy also varies by gender, ethnic group and age. Men receive on average more MTR than women, owing to their higher incomes. Pensioners in particular receive the lowest subsidies and are over-represented amongst the very poorest households (Brownill *et al.* 1990).

A DIVIDED CITY

But London's housing crisis is not a product of housing policy alone. The inter-action of socio-economic change and the housing market has been another vital influence. Demographic factors have had a part to play. Household sizes have decreased while the number of households has increased in the past decade. At the same time after years of decline population levels stabilized in the 1980s and are set to increase to 6.8m. in 1996. As the capital city London offers special attractions to employers and those seeking work. One-person and elderly house-holds are increasing. It is estimated that in 1996 one-third of households in London will be single-person. These trends have all contributed to demand for housing but their translation into housing opportunities depends on the housing finance system and the supply-side response.

As will be seen in the next chapter, global economic shifts were particularly marked in London over the 1980s. Coupled with the economic boom in the South-East in general this led to an unprecedented increase in demand for housing. This boom turned into an equally unprecedented slump when economic conditions changed. Economic restructuring has also brought with it a shift in the composition of the labour market which has interacted with housing market changes.

These changes have seen a rapid growth in employment in the service and financial sectors and a decline in manufacturing and skilled manual work. Some commentators have seen increasing divisions within the economy as a result of technological and economic change. There is debate over whether this can be seen in terms of a dual economy or a more complex patterning. However, what is clear is that there is a gulf between fairly secure, well-paid work in higher socio-economic groups and low-skilled, insecure occupations. As ever these changes have interacted with existing divisions in the population along the lines of class, gender and race.

In housing terms these changes have had various effects. At one level the influx of highly paid financial service workers fuelled house-price inflation and, particularly in Central London led to increased demand for luxury accommo-dation. This was compounded by the demand for company lets, dwellings bought by companies for investment but lived in by their executives or guests in the short term. Central London accommodation became a separate market from the rest of the city with investors choosing whether to buy a Van Gogh or a London flat.

As Smith and Williams (1986) have noted another housing effect of these processes of restructuring is to increase the pressure from affluent workers on residential areas near the city centre. The 'gentrification' of inner-city areas continued in the 1980s and the building of luxury developments became commonplace.

On another level those not in work, those on low incomes or in insecure employment, including the unskilled and 'secondary' workers, black and ethnic minority households, the single, the elderly and women found their access to housing restricted to a shrinking social housing sector. As access to housing has become even more linked to ability to pay, position in the labour market has interacted with housing market position to an even greater degree. Some writers commented on a growing 'socio-tenurial polarisation' (Hamnett and Randolph 1986) as a result of these interactions. They have argued that London was becoming split between an affluent owner-occupied sector and an increasingly residualized population in council housing made up of those on benefits, ethnic minorities and the elderly.

These socio-economic processes are translated into spatial patterns. The contrast between areas in terms of housing conditions, needs and opportunities is now marked in London. Some would argue that this represents a reinforcing of the traditional inner/outer split, with Inner London increasing its concentration of council tenants and the poor and working class. Others see more of an East/West division, and this is certainly borne out in the census and other data, with East London having higher concentrations of lower-class and black residents, council housing, lower housing prices and greater housing needs. However, broad spatial patternings gloss over the fact that contrasts occur at a much finer level. Throughout London pockets of high-value well maintained housing coexist alongside other housing in poor conditions. Whatever spatial patterns are observed the fact is that London is not a homogeneous city and the inequalities between areas are deepening.

Spatial patterns are themselves only a reflection of deeper inequalities that can be found throughout society. From our account so far it has become clear that the housing crisis in London is affecting some sections of the population disproportionately. Existing disadvantage and discrimination is being reflected in the operation of the housing market. The movement towards market provision and the inequitable operation of the housing subsidy system ensures that single people, the low-paid, ethnic minorities, women and the elderly have restricted access to housing, get less financial assistance towards paying their costs and are living in poorer conditions. The present system is helping those that need least help with their housing while painfully failing to provide for those in most need.

PLANNING, HOUSEBUILDING AND THE DEVELOPMENT PROCESS

A final issue that has to be considered in any discussion of the London housing situation is the part played by planning and the development process. As noted above, the belief of government that the market would provide for housing needs has been shown to be false. Why? Ball (1983) and other writers have argued that the British housebuilding industry is driven by the search for profits and is particularly prone to booms and slumps. It is therefore highly speculative, competitive and unstable and does not respond to need. The history of Docklands, for example, clearly shows how private builders were unable to meet the demand for low-cost housing in the area (see Chapter 11 and Brownill 1990). Similarly, the susceptibility to downturns in the market meant that when demand conditions changed from 1989 onwards, supply suffered further.

Within London, particularly central London, the industry comprises property developers rather than the volume builders responsible for the majority of UK housing. These developers are more prepared to take on the risk and the large-scale investment needed. They are even more likely therefore to seek the highest profit rather than respond to demands for low-cost housing.

Property development, as opposed to housebuilding, has also had its impact. Responding to changes in the London economy the speculative property sector has proposed a number of large-scale major developments on the fringes of the city. Called 'mixed-use', they are largely commercial developments with some retail, leisure and residential elements. Docklands, Broadgate, Spitalfields and King's Cross are all examples of existing or proposed developments. These developments threaten the working-class housing in these areas through direct demolition, gentrification and the increased incentive for tenants to buy; land values increase, thereby making the provision of affordable rented housing under present subsidy regimes impossible.

The rôle of planning in relation to housing is often portrayed in a negative fashion. Restrictive policies such as the green belt and the zoning of land for non-residential uses are held responsible for land shortages and high land prices. There can be no doubt that land in London is in short supply. The amount of building on infill sites and back gardens pay witness to this. However, it is a matter of debate whether this is a result of the planning process alone (Evans 1987; Grigson 1986). Often pressure for land occurs in areas of high house prices and steady demand, where builders can make high marginal returns. Other land is left vacant. The London Research Centre has estimated that there is enough land to meet needs into the next century. The question is whether or not builders will build on it.

Conversely, planning has not had sufficient powers to intervene in the land and property markets to the extent needed. For example, zoning land by tenure is not possible under the present system, with the result that land values cannot be controlled.

As a result of the financial restrictions placed on them by government, local authorities and housing associations have been seeking to work more closely with the private sector in recent years. During the boom years up to 1988/9 many planning agreements were struck whereby a certain amount of rented or low-cost housing was provided by developers in return for planning permission, land or other facilities provided by the authorities. After the slump builders were keen to enter such agreements to ensure a certain level of sales. However, this is a way of providing low-cost housing which is highly susceptible to changes in the market. As few new houses are being built and margins tighten for builders, planning gain slips off the agenda.

ALTERNATIVES – WAYS OUT OF THE CRISIS

The complexities and the many facets of the housing problems facing Londoners mean that there is no simple solution. Further, the solution cannot be at a local level alone; national policy change and action is required. But while wholesale reform may be desirable, successive governments have seen it as too drastic and politically damaging to implement.

The scale of the issue is also reflected in the fact that alternatives often come in two time-scales: emergency measures to deal with immediate issues such as homelessness, and long-term reform. The fact that immediate crises cannot be solved without fundamental policy changes is obvious. However, a vicious circle has developed of acknowledging both that problems need deep-seated solutions but also that these may well not come about – so major reform is once again postponed. In the absence of central government reform, proposals have had to focus on particular issues that can be implemented in the immediate future. The Association of London Authorities (ALA), for example, has drawn up a five-point strategy including an emergency programme for London's homeless, help for home-owners who have difficulty meeting mortgage payments, restoring security and protection to private tenants, allowing local authorities to reinvest the proceeds from sales in new and improved dwellings and redirecting more resources towards London in recognition of the capital's special needs (ALA 1990b).

The ALA's emergency homelessness programme aims to bring the crisis under control in five years (ALA 1990a). Aimed at ending the use of temporary accommodation rather than housing all the homeless, the programme involves local authorities and housing associations working together to provide 150,000 new homes. This figure would be achieved through rehabilitation and new building and through taking advantage of the slump in the property market to acquire private property at competitive rates. Importantly, the ALA points out that the cost of this programme would lead to savings in the short term as acquisition is cheaper than temporary accommodation. Yet the programme acknowledges that long-term planned investment in public housing is needed to ensure sufficient housing provisions.

Other solutions, although more long-term, fail to consider the entirety of the problem. Peter Hall (1989) for example sees the issue in terms of a mismatch between dwellings and households and advocates new and expanded towns in the South-East to be developed by public/private partnerships. Such a proposal, however, fails to deal with the issue that owner occupation has patently failed to meet the needs of large numbers of Londoners.

21

Mention should also be made of the many on-going initiatives in the capital which are trying to meet needs and change many aspects of housing provision. Groups involved with co-operative housing and self-build are exploring alternative forms of housing provision. Coin Street and other low-cost inner-city developments are not only providing low-cost housing in the centre of the capital but also placing housing within the overall development of a locality including employment, leisure and retail aspects. Many important ideas are already being practised by such projects. What is needed is to translate the principles for reform which they embody onto a wider scale.

What has also become clear from our discussion in this chapter is that alternatives have to deal with the whole range of factors which are jointly leading to London's housing problems. Dealing with specific factors in isolation will not by themselves solve the problem. Instead of a piecemeal or short-term view a wide-ranging approach has to be taken which covers the provision of both owner-occupied and rented housing, housing finance, taxation and subsidies, equalities issues and participation. We also need to move away from tenure-bound discussions to allow choice and variety and so ensure that housing provision can be flexible in the face of people's changing housing needs (Ball 1986).

While debate on reform of many aspects of housing provision and finance is raging (Ball 1986; *Report of Inquiry into British Housing* 1985) we have space here only to set out some of the most important criteria for reform. It is clear from our review of the capital's housing that the supply of low-cost rented housing has to be increased. Similarly, greater effort has to be made to divert resources to reduce the costs of housing for those that need most help. Disadvantage and discrimination in the housing market has to be tackled and finally housing provision has to be opened up to greater accountability and user control.

Some possible ways in which these goals may be achieved are set out below.

1   Concentrate financial resources on reducing the housing costs of those in greatest need by abolishing MTR or phasing it out by targeting those who need it. To be accompanied by a fundamental reform of Housing Benefit.
2   Increase the supply of low-cost rented accommodation by increasing local authority spending through removing capital receipts restrictions and diverting savings from MTR. Provide more money for housing associations.
3   Reduce the cost of providing social rented housing and therefore rents by giving more planning powers to local authorities so they can zone land for social rented housing. This latter policy has been well argued for by Campaign for Homes in Central London (CHiCL) as essential to providing low-cost housing in the centre of London and where development pressures are bidding up land values.
4   Ensure equalities issues are higher up the agenda. This would involve monitoring the allocation of housing and the operation of private landlords, councils' housing associations and estate agents. Financial assistance would also need to be targeted to specific groups.

5 Increase accountability, user control and participation in the provision and management of housing.

6 Regulation of the private rented sector and registration of all landlords.

7 The rôle of local authorities should be extended to provide a strategic overview across all tenures. New powers are needed to enable authorities to intervene, to draw up strategic plans to meet needs and to co-ordinate activity.

While such a range of reforms might appear daunting there can be no doubt that the scale of the problem demands far-reaching action. We have sketched out the dimensions of the London housing crisis. The use of figures and statistics can blur the reality that these facts are being lived out every day by people on the streets of London, in bed-and-breakfast and in expensive yet inadequate accommodation. The consequences of inaction will be felt by all Londoners, employers and the London economy. This situation can only get worse until radical reforms are implemented.

REFERENCES

Association of London Authorities (ALA) (1990a) *The Homes Front: An Emergency Programme for Housing the Homeless,* London, ALA.

—— (1990b) *Ten Years of London's Housing,* London, ALA.

Ball, M. (1983) *Housing Policy and Economic Power,* London, Methuen.

—— (1986) *Home Ownership: A Suitable Case for Reform,* London, Shelter.

Brownill, S. (1990) *Developing London's Docklands,* London, Paul Chapman.

Brownill, S., Sharp, C., Jones, C. and Merrett, S. (1990) *Housing London: Issues of Finance and Supply,* York, Joseph Rowntree Foundation.

Crane, H. (1990) *Speaking from Experience; Working With Homeless Families,* London, Bayswater Hotel Homelessness Project.

Conway, J. and Kemp, P. (1985) *Bed and Breakfast; Slum Housing of the Eighties,* London, SHAC.

Campaign for Homes in Central London (CHiCL) (1988) *Access to Housing. The Problems for Employers and Potential Employees,* London, CHiCL.

Evans, A. W. (1987) *House Prices and Land Prices in the South East – A Review,* London, House Builders Federation.

Hall, P. (1989) *London 2001,* London, Unwin Hyman.

Grigson, S. (1986) *House Prices in Perspective: A Review of South East Evidence,* London, SERPLAN.

Hamnett, C. and Randolph, B. (1986) 'The role of labour and housing markets in the production of geographical variations in social stratification', in K. Hoggart and E. Kofman (eds), *Politics, Geography and Social Stratification,* London, Croom Helm.

London Housing Unit (1990) *London Needs Homes,* London, LHU.

London Research Centre (1990) *The Demand for Social Rented Housing In London,* London, LRC.

Maclennan, D., Gibb, K. and More, A. (1990) *Paying for Britain's Housing,* York, Joseph Rowntree Foundation.

Report of the Inquiry into British Housing (chaired by the Duke of Edinburgh) (1985), London, National Federation of Housing Associations.

Sharp, C., Jones, C., Brownill, S. and Merrett, S. (1990) *What Londoners Pay for Their Housing*, London, SHAC.

Single Homeless in London (SHIL) (1989) *Report of Working Party*, London, SHIL.

Smith, N. and Williams, P. (1986) *Gentrification and the City*, London, Allen & Unwin.

Threshold (1990) *Hidden Homelessness*, Threshold Housing Advice Annual Report, London, Threshold.

3

EVERY JOB AN OFFICE JOB

Andy Coupland

The past few years have seen a dramatic shift in the nature of London's economy. Twenty years ago there were twice as many workers in London's manufacturing industries as there were working in financial services. Manufacturing industry has steadily declined, while financial services have continued to grow until by 1989 the picture was completely reversed and there were twice as many people working in financial services as remained in manufacturing industry (Figure 3.1).

Alongside this change investment in new office development has soared while investment in new industrial space has fallen. Bank lending to property companies is at a record level, and nearly 200 million square feet of further office space is planned to be built by the end of the decade.

But London's pre-eminence as the European financial capital faces greater challenges as Eastern Europe opens up and Germany reunites. Will all the office space be needed? Has the collapse of manufacturing, and the concentration on office-based service jobs left London with a potentially unstable local economy?

THE BIG SHIFT

London, as the capital city, has a number of characteristics which no other UK city shares. Its potential for tourism, its importance as the seat of government and the home of the Royal Family and most of all its rôle as the headquarters for major UK companies and financial institutions, as well as for European and International bodies and firms, gives it a unique employment profile.

But until a few years ago London was also like any other major conurbation, a place where people made things, imported and exported goods; a city with a diverse local economy.

However, at an ever faster pace London has lost its manufacturing activity, and much related service activity as well. Up to the mid-1950s London had seen consistent growth as a manufacturing centre. Outer London in particular saw the creation of new factories producing 'state of the art' consumer goods, and the forerunners of business parks were established, creating a base for engineering and electronics companies. From the mid-1950s, however, encouraged at the time by government policies which created New Towns, jobs started to dis-

25

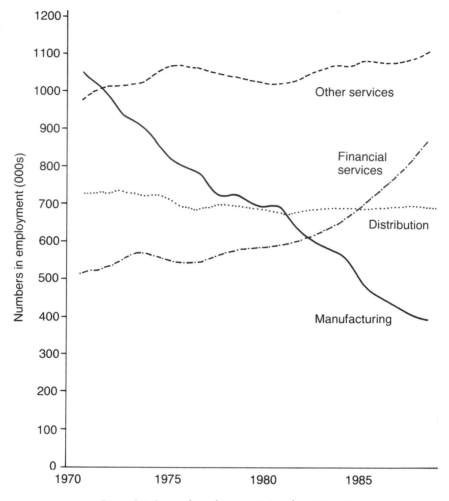

Figure 3.1 Sectoral employment in London 1971–89

Source: London Planning Advisory Committee 1987 and 1990

appear. Some went as companies, unable to expand on their existing sites, moved out, often to areas where financial incentives were available. The merger or takeover of household names led to losses through 'rationalization' while other jobs were lost in the knock-on effect from the closure of the docks and the removal of financial benefits from processing imported raw materials close to the wharves and dockside.

The scale of the collapse in manufacturing has been dramatic. Between 1971 and 1975 19 per cent of manufacturing jobs disappeared. From 1975 to 1981 the fall continued at the same rate – a 19 per cent drop was observed, although most of this took place in the first three years from 1975 to 1978; between 1978 and 1981 some 98,700 jobs in manufacturing and production were lost. By 1981 there was far less manufacturing capacity left in London, and percentage losses slowed in 1981 to 1984, so that only around a 10 per cent fall was observed – but this was a further 142,600 jobs. It was thought that those firms which remained in London occupied a niche which made them more robust, and less likely to be affected by generally adverse economic conditions, but manufacturing here continued to perform less well than in the rest of the country.

Observers throughout the 1980s predicted that the worst was over for London's manufacturing industry, and yet every subsequent set of figures disproved this belief. From 1986 to 1989 a further 81,000 jobs were lost in manufacturing, and there are no signs that this situation is changing today. One explanation is that, nationally, there has been an unexpectedly rapid improvement in productivity in manufacturing. Firms in London, however, are unable to expand production because they cannot obtain land to build on; hence rising productivity in general means London's manufacturers become less competitive and so close down or move away. By 1989 only 12 per cent of London's jobs were in manufacturing.

Current projections from the London Planning Advisory Committee (LPAC) are that further falls can be expected; their 1987 projections (PA Cambridge Economic Consultants et al. 1988) were reviewed in 1990, and now a faster rate of decline is forecast. By 2006, the estimates imply that London will have only 5 per cent of its workforce employed in manufacturing industries. Figures from another projection by Roger Tym & Partners (1990) show almost identical results.

At the same time that manufacturing started to decline seriously, throughout the 1960s and 1970s Britain's position as a centre of international financial trade grew stronger. However, fewer new jobs were created in this area than might be expected as 'back office' service jobs, particularly in areas like insurance, were moved out of London. The population of the city continued to fall, so some service jobs which relate specifically to the number of people living in the capital fell too. This was offset by the rising standard of living which led to a demand for a wider range of services.

Up to the mid-1980s London's economy had taken a smaller share of Britain's total economic activity every year for over twenty years. From 1977 however, there was a marked shift, with London's performance closely matching (albeit slightly worse than) that of the rest of Britain. Up to the late 1970s London was subject to specific controls on development for both offices and industry. The worsening performance of London led to a review of this situation, followed a few years later by the creation of specific policies to overcome some of the collapse of the inner-city economies of Britain – including London's. Experiments were created to introduce both an Urban Development Corporation and an Enterprise Zone into London's east end – an area where the local economy had collapsed more than most, in part owing to the closure of London's upstream docks.

Office Development permits had been scrapped some years earlier, and developers responded by buying sites in and around the City to develop a new type of office space. Until the early 1980s the property market was effectively structured to meet the demands of the traditional financial institutions, which in turn provided or arranged the financing of the buildings.

Location was the crucial factor for the financial institutions – the closer to the Bank of England, the happier the companies were. Then in the late 1970s and early 1980s American and Japanese banks started to increase their rôle in the UK. Firms like Citicorp arrived in strength, prepared to compete with the High Street banks, and willing to apply different criteria to their space requirements. While they needed to stay in London for business reasons and so that they could obtain qualified staff easily, the new arrivals were less concerned about exactly where they were located. US banks took buildings in the West End, in Covent Garden, in Victoria, and Citibank even went south of the River Thames to London Bridge City. By 1988 there were 20 more US banks based in London than there were in New York.

From the mid-1980s onwards there was an even greater increase in the numbers employed in financial and 'city' jobs. In part this expansion in employment is explained by the 'Big Bang' of 27 October 1986, when the City reorganized itself, allowing foreign-owned companies to trade in stocks and shares. At the same time traditional barriers between jobbers – who sold shares as wholesalers – and brokers – who bought them for institutions or individuals – were broken down.

As part of the lead-up to the Big Bang, European, American and Japanese firms all expanded their rôle and activity in London. Some set up their own trading offices. Others chose to buy existing broking firms. Six of the biggest ten brokers in London became foreign-owned. UK banks also bought broking firms, and often merged several smaller companies, each with a specialism within the market. The new firms all invested heavily in office space, with dealing floors forming an integral part of their operation.

Each firm now carries out the share-dealing operations in its own buildings – none are left on the floor of the stock exchange itself. This led to a dramatic upsurge in demand for office space – particularly that space which could incorporate the large column-free areas of dealing space.

Banks replaced pension funds and other institutions as the main source of development finance during the 1980s. At first the funds increased steadily; but by 1987 banks were investing over £13bn in commercial property. More was lent in the first three months of that year than during the whole of 1985. Syndicated, that is jointly financed, loans raised over £3.3bn in 1988 – compared with £91m. in 1983. Despite warnings from the Bank of England of the potentially dire consequences of continued loans to property companies, the money still pours out. By 1990 bank lending to property companies had risen to £38.9bn, and the total exposure of banks was over £450bn. Nearly half of the money lent in 1990 came from banks based outside the UK. As property companies steadily went into receivership at the end of 1990, a survey by Woolgate Property Finance discovered that only 18 per cent of banks were expecting to reduce their property lending, while 38 per cent expected to increase it (reported in the *Estates Times* of 30 November 1990). Clearly, then, the financial institutions putting the money up – increasingly overseas banks and particularly the Japanese – saw no reason to panic about the state of London's property market. Yet expectations about the continued growth in employment in the sectors which occupy the new office buildings may be disappointed, and there are signs that the scale of growth of new buildings will not be matched by an equivalent growth in demand for that space.

By June 1991 vacant office space in the City of London had soared to 18 per cent, depressing rents and capital values. Meanwhile demand was slumping as supply increased further, and completions in fringe locations complicated the picture further. Docklands saw figures of 40 per cent vacancy rates, and rent levels halved to only £10 and £12 per square foot (a quarter of the equivalent City rental level).

The problem with these shifts is that London, with no obvious planned intention, has ended up with a highly volatile and unbalanced economy. Many local services are either no longer obtainable, or very expensive to provide as land values rise and manufacturing and low-value service jobs disappear. The growth of office development has pushed expected land values to remarkable levels, ensuring that a diversity of employment opportunities is no longer available. With limited training or retraining opportunities, the poorly qualified and unemployed in London have little chance of getting a job. A collapse in the financial services sector would see London's economy in a very bad way. In the meantime the concentration of employment in the City, the fringes of the City and the newly created east end financial centre on the Isle of Dogs is creating a serious problem for London's transport infrastructure, and requiring massive public sector investment to upgrade the system.

SPATIAL IMPLICATIONS WITHIN LONDON

In recent years we have come to think of 'the inner city' as a problem. London's 'inner City' has indeed experienced the greatest increases in unemployment, but London is not quite as straightforward as that. The greatest increases in new jobs have been registered in Central London, although so too have the greatest number of job losses, particularly in East London.

Overall London lost jobs throughout the 1970s and for the first half of the 1980s. Between 1971 and 1981 there was a 12.5 per cent drop in the number of employed residents in London – 455,000 people. But Inner London lost the vast majority of these residents: 363,000 or 18 per cent of the Inner London total, with just 93,000 coming from Outer London, a 5 per cent fall for that part of the capital. Between 1971 and 1981 Inner London residents lost jobs in Inner London at a faster rate than jobs declined in the Inner areas. While Outer London residents appeared to benefit from this situation, with females in particular increasing their take-up of jobs in Inner London, they lost out in Outer London to the rest of the South-East (ROSE) residents.

However, overall there seems to be a greater similarity between Inner and Outer London's economies. Recent trends in rising unemployment during 1990 have focused more on Outer London (and interestingly West London) than Inner (and East) London.

While distinctions between Inner and Outer London are perhaps becoming less obvious, those between East and West London are still very apparent. While percentage employment declines between Inner and Outer London were roughly the same between 1981 and 1984, East London lost 11.5 per cent of jobs, while West London lost only 1 per cent. If anything this trend has become more pronounced over the past decade.

This follows a trend established back in 1971. Tables 3.1 and 3.2 show that in West London there were net losses of employment, but from 1971 to 1984 the

Table 3.1 Change in manufacturing jobs in London

	1971–8	*1978–81*	*1981–4*
Central London	−49,405	+ 4,032	− 3,842
Western fringe	−26,172	+ 2,845	− 5,503
Inner West – remainder	− 9,489	− 3,937	− 3,902
Outer West	−72,021	−22,186	−37,986
Eastern fringe	−28,691	− 7,458	− 8,438
Inner East – remainder	−46,538	−38,071	−24,084
Outer East	−25,587	−25,777	−15,292

Table 3.2 Change in service jobs in London

	1971–8	*1978–81*	*1981–4*
Central London	−61,084	− 9,108	+33,926
Western fringe	−18,143	−12,265	+ 8,862
Inner West – remainder	+13,413	+ 224	− 5,978
Outer West	+68,229	+16,171	+23,357
Eastern fringe	−12,775	+ 7,760	+ 5,892
Inner East – remainder	+24,751	−28,195	+ 5,122
Outer East	+22,903	− 3,206	− 90

Source: Compiled from Employment Topic Working Party, LPAC, 1987

recorded losses in manufacturing industry of over 175,000 jobs were offset to a great extent by the 90,000+ additional service jobs. In East London however, 220,000 jobs were lost in manufacturing but only 20,000 additional jobs were created in service industries. In future this situation is expected to change. The greatest predicted employment increases are expected in Inner East London; not least because of the massive employment growth which is expected to be associated with the development programme in London's Docklands, which could see 110,000 more jobs in that area by the year 2001.

Table 3.3 shows that of the extra jobs in services in West London, two thirds of these can be accounted for by business, finance and related services, while East London saw 45,000 new jobs in this category, but 25,000 fewer service jobs in other categories. The most recent period for which figures are available confirms the trend for the West to prosper, with 1 per cent employment growth in the City and Westminster, 2 per cent in the inner West and 3 per cent in the outer, while the inner East saw a 1 per cent fall and the outer East a 3 per cent fall.

In terms of resident populations, 21.7 per cent of economically active males were categorized as professional, managerial or intermediate non-manual workers in 1981 for inner East London, while the comparative figure for the inner West was 33.4 per cent. For Outer London the figures were 28.2 per cent in the East, and 38.2 per cent in the West. It seems clear that East London has a higher proportion of manual and unskilled workers, and that Inner London has a higher proportion than Outer London.

LONDON AND BEYOND

A majority of the people who live in London have jobs in the capital. Figures are only available from the 1971 and 1981 censuses but these show wide variations between boroughs, and a slight decrease in the percentages for most boroughs in the proportion of workers that are employed locally. At the top of the percentage

Table 3.3 Change in business and finance service jobs in London

	1971–8	1978–81	1981–4
Central London	−11,675	+13,580	+29,674
Western fringe	+ 1,100	− 4,629	+ 5,197
Inner West – remainder	+ 7,701	− 1,115	+ 1,696
Outer West	+21,470	+ 5,229	+24,179
Eastern fringe	+ 2,867	+10,850	+ 1,275
Inner East – remainder	+13,069	+ 6,456	+ 4,876
Outer East	+ 5,852	+ 973	+ 2,807

Source: Compiled from Employment Topic Working Party, LPAC, 1987

31

figures, 68 per cent of Westminster residents worked locally in 1971, but by 1981 this had fallen to 65 per cent. Of Kingston's residents 57 per cent had local employment in 1971, but by 1981 only 50 per cent were in this category. At the bottom end of the scale only 37 per cent of Lewisham's residents had local jobs in 1971 and 1981.

Increases in the numbers employed locally were seen in only five boroughs including Southwark, which increased from 46 per cent to 50 per cent, Camden, which rose 1 per cent to 48 per cent, and Kensington and Chelsea which rose to 41 per cent from 37 per cent. Overall, London residents were losing out to those from outside the capital. Part of this can be accounted for by the fact that there was an 18 per cent fall in the resident labour force between 1971 and 1981, but this in itself does not fully explain the fact that London residents' share of jobs dropped from 86 per cent to 82 per cent.

A similar change can be seen if the percentage of residents in the local work-force is studied. Again, in just five boroughs the percentages increased between 1971 and 1981: in Hillingdon, Ealing, Croydon, Harrow and Waltham Forest, and in none of these by more than 2 per cent. Everywhere else stayed the same or fell – in Southwark from 41 per cent to 32 per cent, in Lewisham from 52 per cent to 47 per cent and in Bromley from 70 per cent to 66 per cent.

These figures show a general trend. Over the past few decades more and more of London's jobs have been taken by workers travelling into the capital – the vast majority from the rest of the South-East (ROSE). By 1981 18 per cent of London's workers, 31 per cent of those in the City of London, came from ROSE. Between 1971 and 1981 the percentage increases of ROSE residents working in London increased substantially, particularly for female workers whose rate of increase was 32 per cent in inner East London and 28 per cent in inner West. These figures are even more significant when the fact that the number of jobs in London for this period fell by 12 per cent is taken into account.

It is highly unlikely that these trends have changed very much in the past decade, or will do so in the near future. The only factor which may change the situation is if jobs move even faster from London into the rest of the South-East. The economy in ROSE has already been boosted by such moves: between 1971 and 1984 198,000 manufacturing jobs were lost, but 647,000 new service jobs were created, an increase of 34 per cent. By 1985 there were over two and a half million service workers, compared to under a million in manufacturing. (In the same year in London there were a similar number of service workers but under half a million in manufacturing.)

Relocations out of London are still significant; in 1990 34 companies left the capital with 11,380 jobs moving, an increase of 80 per cent on the total for 1989. However, this may not benefit the rest of the South-East; of the 14,500 jobs which were expected to move in 1991, over half were to areas outside the South-East (Jones Lang Wooton 1991).

The steady shift to service employment, and the increasing commuting to these office-based jobs has huge implications for the transport network. The planned growth of London eastwards, and the careful integration of the new employment opportunities into the changing local economies might have been accomplished relatively painlessly. However, in a vacuum of strategic and regional investment, and with huge changes forced onto East London through the imposition of a Development Corporation and the creation of an Enterprise Zone, chaos reigns. Londoners continue to lose out, the transport network slowly grinds to a halt and fewer traditional employment opportunities are left for the unskilled, or those whose skills no longer match the jobs on offer.

THE FUTURE

So what is the future for London's economy? Manufacturing is disappearing fast – indeed so much has already gone that whatever happens to that which is left, it can hardly affect the overall figures. The main sector which matters is that of finance, banking and related businesses. Here too decline is forecast, but potentially with far more severe consequences. The Cambridge Regional Economic Review (1989) predicts that London's slight increase in jobs will reverse to an average fall of 1 per cent throughout the 1990s. Already London is less important than it once was in Britain – in the mid-1950s it had 20 per cent of the entire UK employment market, but by the end of the 1970s this had fallen to under 15 per cent. Despite this London was still a conurbation with a better employment situation than any other in the UK; this may soon change. The Cambridge report expects that, with continuing development and relocation of government and office functions both Greater Manchester and West Yorkshire could be in a better position than London by 2001.

However, other forecasters are more hopeful. The Kent University/LPAC forecasts produced in 1990 show continued growth of 340,000 jobs by 2006. This is in part because 1986–9 saw an unexpectedly large increase of 155,000 jobs. The LPAC forecast assumes that this rate of growth will not be sustained. Nevertheless, if the assumptions are reasonably correct over two-thirds of all London employees will be involved in finance, business or the public sector (which is not expected to grow during the period). Manufacturing industry is expected to lose a further 169,000 jobs during this period, leaving it with around 5 per cent of all the jobs.

This part of the forecast is one which few would argue with. Effectively no new industrial premises are being constructed in London. The changes to planning legislation which grouped together office uses, research and development and light industry in the 'B1' business class means that permission is usually not needed to change from light industry to office use. Other legislation allowed general industry to change to B1, and hence offices without the need for planning permission. These changes have led to an acceleration in the loss of industrial floorspace, particularly in Inner London boroughs. Now Outer London, and much of the rest of the country, but particularly the South-East and motorway corridors out of London are seeing the creation of huge 'business parks', which are effectively collections of office or closely allied functions provided in a lower-density landscaped campus. Many of these are being built on former industrial sites.

Questions must remain about the scale of the continued increase in the finance and business sectors. Despite the bankers' expectations of continued lending to property developers, there is a vast oversupply of office space in London, which in turn is pushing rents lower. These lower rents may, ironically, help London maintain its competitive edge against other provincial or European centres, but there are doubts whether property companies can sustain lower rents – particularly if interest rates remain high through 1991 or beyond. Many companies are already stretched, some are overstretched, and a substantial collapse of the property market would have a devastating effect on London's money markets and the attitude of many overseas institutions, which are often the same firms that have lent the money to create the oversupply.

There are signs that, after many years of authors predicting just such a development, companies are finally taking the advantages offered by new technologies and moving out of London. Relocations of the 'back office' functions of insurance companies date back to the 1960s. In recent years Pearl Assurance moved their entire UK Headquarters to Peterborough, Sun Alliance to Bristol and the Trustee Savings Bank to Scotland. Many government clerical functions have followed this trend, and more are planned. Now firms are taking their main headquarters too; Lloyds Bank have completed the first phase of their new buildings in Bristol, with more to follow in 1992.

The greatest unknown, however, concerns the firms recently arrived in Britain, and those which have yet to decide where to locate. The constant government advertising about the need to prepare for the common European market in 1992 shows the importance which it carries. For both US and Japanese companies Britain is one of the more obvious choices for European headquarters. Language barriers are less of a problem. The stock exchange is accessible and London still has a unique package of financial institutions offering every business service that large multinational companies need. There is still land available for new development, although mostly on the fringes of the City or in western centres such as Hammersmith, and there are new schemes completed, under construction or with approval that can be tailored to a client's needs.

But there are signs that some companies are thinking beyond these advantages. London is an expensive city to locate in – only Tokyo has rents that are more expensive than those of prime office space in the City of London. Only a few companies are rethinking their decisions to come to London, but they may be the first of many. One US bank has already decided to relocate its European headquarters to Frankfurt. Equally significant may be the decision at the end of 1990 by a major Korean electronics manufacturer, Samsung, to pull out at the last minute from a decision to locate their European headquarters on the edge of London in Brentford and instead, also, to locate in Frankfurt.

The decision to move to a more easterly located European city starts to make

34

sense when the reunification of Germany and the democratization of Eastern Europe are taken into account. There is a tremendous interest among office users in taking up space in Frankfurt and Berlin, and to a lesser extent Paris. This is reflected in the increase in rents in these centres. While office rents in London have been stable or fallen in 1989 and 1990, in Frankfurt they rose by 14.3 per cent in six months to December 1990, and in Berlin by 33.3 per cent. UK developers like London & Edinburgh Trust are planning schemes in Berlin, and a boom is predicted for the city over the next few years, although this may be slow, given the need to develop new infrastructure.

This situation places even greater pressure on London, where oversupply is now a notable factor. London agents Baker Harris Saunders say that this is likely to mean no increase in office rents in the city in the next five years. Another firm of agents, Drivers Jonas, preparing a report for London Transport, say that none of the 6 million square feet of space planned in London's Isle of Dogs Enterprise Zone is likely to be started in the next five years. To quote a company spokesman: 'Central London will remain firm but the peripheries will be blown out of the water' (*Estates Times*, 18 January 1991).

The oversupply of space must have direct connections to the employment situation. However, the indications are that the property market may have been over-optimistic. Roger Tym & Partners, in a study for London Underground, London Transport, Network SouthEast and the Department of Transport (1990) estimate that the amount of commercial space constructed since 1987, under construction or planned (totalling some 183 million square feet) would be occupied by 620,000 people. Yet no employment projection expects that many jobs to be generated in London. LPAC's projections (1990) expect there to be some 344,000 more jobs in the banking, insurance, finance and other business sectors. However, no other office-using sector of employment is expected to increase. The Roger Tym report predicts only 214,000 more jobs in the business and finance sector, but an increase of 158,000 jobs in public administration, health, education and other services. A report from Cambridge Econometrics/ NIERC (1990) predicts a fall in London's employment base by 72,000.

Clearly no projection is likely to be an accurate prediction. The likely state of affairs is that London's economy may stabilize at present levels, with a further modest growth in support activities such as accountancy, legal services etc., little or no growth in the finance sector, a move out from London by many support services of government and a further severe fall in the percentage of manufacturing jobs (but quite a modest number given the few jobs now existing in this sector).

If this is a reasonable guess at the next decade, virtually none of the office space now under construction will be needed. If it is occupied, for financial reasons, or because of location factors, or because it represents better-quality space or to centralize a number of scattered offices, it will merely liberate equivalent space somewhere else in London. The future for property companies looks very unhappy. If other factors come into play, and London becomes less favoured than other European centres, or if US companies pull back from Europe, then London, with the majority of its employment eggs in one finance-related basket looks to be set for a very difficult, and potentially disastrous decade.

REFERENCES

Cambridge Econometrics / NIERC (1990) *Employment Forecasts*, Cambridge, Cambridge Econometrics.

Cambridge Regional Economic Review (1989) Department of Land Economics, Cambridge, University of Cambridge.

Hall, P. (1989) *London 2001*, London, Unwin Hyman.

Henley Centre (1990) *London 2000*, report produced for the Association of Local Authorities by the Henley Centre, Henley.

Jones Lang Wooton (1991) *The Decentralisation of Offices from Central London*, London, JLW.

London Planning Advisory Committee (LPAC) (1987) *Employment – Report of the Topic Working Party*, London, LPAC.

—— (1990) *Employment Projections for London*, London, LPAC.

PA Cambridge Economic Consultants & Urban and Regional Studies Unit of the University of Kent (1988) *Economic and Employment Change in London*, Cambridge, PA Cambridge Economic Consultants.

Roger Tym & Partners (1990) *South East Employment 2001*, London, Roger Tym & Partners.

4

JAM TODAY: LONDON'S TRANSPORT IN CRISIS

Ruth Bashall and Gavin Smith

Nobody says 'Crisis? What crisis?' about London's transport. People of all political persuasions can only agree that London's transport is in a mess. Without transport – in any of its manifestations including telecommunications, pipelines, postal services, door-to-door deliveries, as well as the more obvious buses, trains and cars no part of the city's life can operate.

In other sectors of public services and industry, advantaged customers can buy into exclusive private systems that circumvent the standard public ones, but this is possible only to a slight degree in the transport sector. True, the wealthier business person can be flown by helicopter, and the Ministry of Defence can capture its own radio frequencies; but most users of the space required for movement immediately bump up against all the other people doing the same thing. A Rolls Royce in a traffic jam in the Kings Road goes no faster – in fact, slower – than a bicycle on the same street. Water mains serve Chelsea as well as Paddington. Most transport necessarily utilizes public infrastructure open to all.

Current government policy fails to grasp this nettle. One Transport Minister's dismissal of the problem, on the grounds that 'increased traffic' – for which read congestion – 'is a sign of economic health' has had to be quietly played down. Vehicle speeds in Central London are falling and at 10 m.p.h. are currently little faster than the horse and cart of the turn of the century (*Guardian* 5 April 1991). Nearly 20 per cent of rush-hour trains are at least five minutes late (Network SouthEast 1990). The London Underground is painfully overcrowded (a 29 per cent increase in passengers over the period 1980 to 1990) whilst the number of bus passengers has declined (Department of Transport 1990). Pedestrian fatalities on London's roads (244 in 1989) are scandalous; as are those of cyclists and motorcyclists. The last few years have seen a series of appalling public transport accidents: 35 dead in the King's Cross Underground fire of 1987; 56 dead in the 1988 Clapham rail crash; 5 dead in the 1989 Purley rail crash; 2 dead and 500 injured in the Cannon Street rail crash of 1991. Not a happy catalogue.

The 'crisis' of transport is in reality a multiple one. The problems of congestion, performance and safety are symptomatic of deeper crises. Pundits, transport buffs and environmentalists agree on this, though frequently disagreeing on the possible solutions. Are London's transport problems to be solved by building more transport networks? By switching from car to train? By transferring to different media (working at home, telecommunications, door-to-door deliveries)? Or should we travel less, consume less, be served by less centralized land uses? Should, for example, the costs of motoring be deliberately raised? Does everyone have an equal right to mobility?

Yes, all this and more, no doubt. Yet to see why so little success has been achieved, we need to examine the deeper problems. These, we suggest, are the crises of investment, organization, marketing and equity. Each must be solved before progress can be anticipated.

INVESTMENT

There isn't enough of it. With the end of the Thatcher era, it need no longer be a point of pride that London's public transport fares are subsidized to the tune of 22 per cent, compared with Rotterdam's 83 per cent, Frankfurt's 55 per cent, Rome's 76 per cent (Bushell 1989).[1] For 'subsidized' read 'supported'. London woefully underinvests, both in its operating and its capital accounts.

A classic case is the continuing lack of a national plan for rail links to the Channel Tunnel and its associated London termini. Many business and independent observers have argued for a joint freight–passenger link to a new Stratford terminal on the edge of Docklands, yet official government policy still offers British Rail a zero budget and continues to fumble with ageing inner-city links through South London that are highly unpopular with residents. London cannot compete in Europe if it is starved of infrastructure.

This collapse of will is not unique. It has been estimated that millions of £s have to be spent on replacing ageing Victorian sewers and water mains throughout London: last year the borough of Camden's entire highways maintenance budget disappeared into three burst water mains (they had been compacted by lorries). Even where the need for a particular expenditure has been agreed, there are debilitating delays. The 'City Commuter Services Group', a group of City interests, argued that investment in commuter rail services to Central London must increase. The result, the government-initiated Central London Rail Study (CLRS) which came out in 1989, proposed a sensible investment programme of new subsurface cross-town links. (Department of Transport *et al.* 1989). As yet no budget has been allocated.

There is cheeseparing of essential maintenance and replacement work. Unsafe cutbacks in cleaning schedules are known to have been an important contributory factor in the King's Cross Underground fire (Fennell 1988). A tragedy waiting in the wings is London Underground's cessation of preventative axle maintenance and its move to 'after the event' axle repairs. Aged rolling stock is believed to have contributed to the high number of injuries at Cannon Street.

One side-effect of the opening of provincial bus services to 'free competition' (or deregulation, as it is known), and in London, the requirement that London Transport tender out contracts to independent bus companies, is over-enthusiastic cost-cutting. Money is not being spent on replacing ageing vehicles. The collapse of the British bus-building industry since deregulation was to be expected and has indeed largely occurred. Paucity of new buses means a paucity of response to new passenger demands (for example for vehicles accessible to disabled people, or for more luggage and storage space) and a downward spiral of consumer satisfaction. CVE, manufacturers of the UK's most accessible minibus, the Omni, went into receivership in 1990.

Underinvestment in the workforce is a malaise normally overlooked. London Buses Limited (or rather, its subsidiaries which operate as semi-private separate entities) actually cut wage rates while competing for London Transport contracts; hours have increased. Employees of London Buses fought hard for a 38-hour week but even staff working for unionized private operators such as Kentish Bus work a basic 42-hour week. Such changes have heralded further alienation amongst London's bus workers, who now display high stress levels, especially one-person-only drivers. Staff turnover is high; passenger empathy, some would argue, is decreasing day by day. Even in these times of high unemployment, both buses and railways are experiencing staff shortages. The industry is unattractive to potential workers.

The managerial response, and a sadly inappropriate one, has been to retain grossly unsocial shift systems and encourage overtime, and sometimes danger-ously long hours. The signals engineer whose faulty work was found to be one of the causes of the Clapham rail crash worked at hourly rates well below those in the private sector, and had had only one day off in three months (Hidden 1989). Not much has changed since; in November 1990, the National Union of Rail, Maritime and Transport Workers revealed that 30 per cent of BR's signals workers are still exceeding even the 6-day, 72-hour week stipulated as a permiss-ible maximum by Sir Anthony Hidden's report into the crash.

Misinvestment is almost as great a problem, and takes many forms. A few examples follow.

'Labour-saving' technology all too often is simply a means of cutting the wages bill whilst taking advantage of random capital grants. London Under-ground spent £145m. installing the Underground Ticketing System, a system of ticket automation with universally unpopular automatic ticket barriers. Such barriers would be redundant under a European-type ticketing system. Further money has had to be spent since, to alter the barriers so that they do effectively lock open in case of emergency.

London Underground and British Rail's response to heightened passenger concerns about security (sexual assaults in 1990 increased 61 per cent on British Rail and 41 per cent on the Underground (British Transport Police 1990)) at unstaffed or understaffed stations and on trains, has been the 'technical fix' of the video monitor; without the staff to monitor and to act on what they see, it is useless. In February 1991, London Underground announced that it was to cut 1,000 jobs to balance the books. What price people's lives?

Company cars receive tax relief; surveys indicate that three-quarters of cars entering Central London are company-supported and that three-quarters of these remained parked all day (TEST 1984). They are a perk, not a business tool. The Department of Transport's thinking reinforces the use of cars for commuting. Current Department policies include: an encouragement of privately funded (toll) roads; 'driver guidance' systems which will allow some car commuters to bypass bottlenecks with the aid of an on-board computer terminal; the declaring of London's radial routes to be through-way 'Red Routes', regardless of the impact of the resultant increase in the speed of through traffic on many of London's local shopping and residential areas.

By contrast, *social assets are drastically underpriced.* Dr John Adams of University College London has pointed out that the Department of Transport, in planning the route of new highway schemes, values public open space negatively, instead of costing an important environmental loss. False pricing is endemic: the Dept of Transport's Cost Benefit Analysis (COBA) of road schemes is used to justify new roads; this on the basis that they generate surplus value by saving accidents and increasing vehicle speeds. COBA fails to cost the effect of the increased traffic generated by new highway capacity, and thereafter the increased road accidents occurring across a wider area of London. Highway investment is counted as a 'net benefit'. The logical alternative option for public expenditure, investment in public transport, is accounted to be a 'cost'. COBA is not applied to bus or rail schemes; instead rail schemes are required to generate an 8 per cent return on investment within the first year. The government specifically disallows a unified approach to the costing of transport options. In economists' terms, the government ignores the issue of 'opportunity costs'.

London's transport policy is very far from being 'green'. Acid smog and traffic noise are visibly and audibly on the increase (LBA 1990). We have witnessed a belated success in reducing traffic-associated lead levels. The lack of a 'green' mentality makes various issues unnecessarily hard to solve. 'Rat-runs' through residential areas, illegal car parking, poor town-centre improvements, and out-of-town car-based hypermarkets proliferate; awareness of the need for priority towards pedestrians and cyclists continues to be low.

Given the above factors, no sensibly integrated investment policy is possible. Interestingly, the raising of private cash through what is known as 'planning gain' has also been problematic. Thus Olympia & York, the Canadian developers of the Canary Wharf office site in Docklands, have provided £68m. towards the cost of extending the Docklands Light Railway to Bank. They also effectively diverted the government's attention from the Central London Rail Study's preferred Paddington to Liverpool Street 'Crossrail' scheme by offering money towards a far lower priority scheme, a Jubilee Line extension through Canary Wharf. The government took the bait, deferred Crossrail, and declared for the Jubilee Line. Yet in the event it appears that less than a third of the money for the Jubilee Line scheme is likely to be from private sources (contrasting with the 60 per cent envisaged) (Olsberg 1990). Research suggests that very rarely does planning gain pass on to the public much of the developers' land value gains associated with publicly financed transport improvements.

The lion's share of investment continues to revolve around highway schemes. The government used private consultants rather than partnerships with the London boroughs to undertake the London Assessment Studies. Predictably, these came out in favour of the very highway schemes that had been in the Department of Transport's bottom drawer for many years. Massive public opposition was foreseen but could not in the end be circumvented (especially as the areas affected included several vulnerable Tory marginal seats); the Minister abandoned the schemes in 1990 – after spending £8.5m. on consultancy fees. Public transport investment schemes continue to be delayed, while roads such as the North Circular and the A13 in Docklands are surreptitiously being upgraded by means of massive interconnecting 'junction improvements' – at high cost to the surrounding communities.

ORGANIZATION

The control of London's transport system is highly fragmented. Its more than 100 miles of trunk roads are managed by the Department of Transport, who also hold a traffic management veto over another 1,000 miles of 'designated' borough roads. British Rail is a quango running one-half of London's railway network; London Underground runs the other half. The government holds the purse strings of London Transport, set up in its current form by the 1984 London Regional Transport Act. London Transport is now merely a holding company for London Underground and London Buses Ltd and has a rather ill-defined planning rôle. In 1991, London Buses, or rather its subsidiary companies, was running 60 per cent of London's bus routes directly; the remainder of the capital's bus services are run under contract by London Buses subsidiaries (who on occasion bid against each other for contracts) or its privately owned rivals, many of them ex-subsidiaries of the National Bus Company, broken up and sold off under the 1985 Transport Act. The government has stated its intention to extend 'deregulation' (unbridled privatization and competition) to London if it wins the next election – though it is hard not to wonder why, if deregulation is as beneficial to the passenger as the government has claimed, London was not the first to enjoy its delights?

Much of the remainder of London's transport systems is effectively controlled by the 33 London boroughs: local roads, pedestrian facilities, cycle routes and the land-use and development control constraints imposed on public and private development by the boroughs as planning authorities.

Transport planning should be intimately related to *land-use planning*. The limited co-ordination that does occur is largely orchestrated by the more progressive boroughs on their own patch, and by consortia of local and regional authorities (the London Planning Advisory Committee and SERPLAN and the two associations of London boroughs). But transport and land-use co-ordination is not the norm, given the fragmentation of transport planning and operation and the power of the Departments of Transport and of the Environment to override the boroughs. Boroughs are left to attempt transport and environmental planning through Unitary Development Plans and their annual Transport Policies and Programme (TPP); government advice is at best complacent, at worst obstructive.[2]

London's transport system has very real problems of *accountability*. The tip of this particular iceberg is secrecy. Since the demise of the Greater London Council in 1986, there has been no democratic control of the system; investment decisions cannot now be derived from popular demands, but are imposed in mysterious ways. Who, for instance asked the Department of Transport to spend so much energy and taxpayers' money researching and trying to implement driver guidance systems? Who asked British Rail to miss out Brockley (the local station of one of this book's contributors) from the schedules of most of the trains that pass through it, no doubt carrying more valuable longer-distance passengers? A tale worth telling is that of Network SouthEast's unilateral decision, as part of its 'station improvement programme', to paint all station fixtures bright red – in spite of protests from rail unions that the red paint would distract drivers from the danger code for signals – and with a paint whose lead content flaunted British Rail's own standards. The explanation is that that year red was the fashion colour for executives' scarves and socks.

MARKETING AND EQUITY

London is no different from other UK cities in that transport decisions are made by white, able-bodied, middle-class males, for white, able-bodied, middle-class males. The problems associated with this approach become obvious when one looks at the characteristics of the pedestrians on the average London street, or the passengers on the proverbial Clapham omnibus: perhaps 60 per cent women, 20 per cent children and young people, 20 per cent men, and significant proportions of black and ethnic minority people and pensioners. This of course does not account for the 'hidden' passengers: disabled people barred from using mainstream 'public' transport by poor design and discriminatory attitudes; women whose journeys do not 'fit' the established bus network, for whom the care of young children renders many forms of transport difficult or impossible, and for whom the risk of attack is a major deterrent to going out; the increasing numbers of people too poor to afford rising transport costs.

The marketing of transport cannot be discussed separately from the issue of accessibility, or equity – in other words, who has access to any particular transport system, and who does not. There are a number of parameters: geographical location, income, gender (and domestic responsibilities), race, disability, age.

Transport operators are fond of quoting 'market research' findings which show that 'demand for a bus service has fallen by x per cent over 4 years' and therefore the service is no longer required. Unfortunately such statements start from a false premise, namely that *demand* is the same as *need* – that if people no longer use a service, it is simply because they no longer need it. This may seem obvious – but let us look beneath the surface: that bus service that had been losing passengers – had it been so run-down that no-one bothered to wait any more? Had several buses been cut out at the time when children come out of school? Had it been converted to one-person operation? Was it still fulfilling the need that it was set out to answer, or had that need altered slightly – for example, perhaps the time-table had not been revised to reflect altered starting-time for a major local employer?

As concepts of market-led demand have replaced the old ideas of passenger transport as a public service, inequalities have increased at all levels. Some inequalities, however, predate current government policies, reflecting both a failure of transport systems and of the wider society to address discrimination.

Geographical disparities in transport provision are apparent across London: for example, in what we might call the 'Brockley station syndrome' already mentioned. British Rail does badly by almost all of Inner London – ludicrously badly in those areas of South London or in boroughs such as Hackney where there is no overlap with London Underground. Bus networks are also unevenly distri-buted: there has been a historical emphasis on longer-distance, radial routes into the centre, with only thin coverage in outer areas. Frequencies of less than 20 minutes are unusual on most outer London routes, with the exception of the newly introduced local 'Hoppa' services. Inner-city residents remain largely dependent on bus services. Residents of new or relatively new estates may find themselves with little or no transport. Transport services have failed to respond to changing patterns of employment, leisure and education. Car ownership is subject to local variation – the South-East has the highest proportion of car-owners in the UK; but whilst the figures for Outer London are high, with many households owning two cars, residents of London's poorer boroughs (Tower Hamlets, Newham, Hackney, Southwark), have some of the lowest percentages of car ownership in the country and suffer jams of through-traffic almost all day long.

The last Greater London Council administration's efforts to introduce some degree of London-wide equity were brought to a halt by the abolition of London's government in 1986. For example, the London Taxicard concessionary fares scheme for disabled people, initiated by the Greater London Council, is now subject to geographical variations and the vagaries of borough politics and finance: Taxicard members in some London boroughs benefit from unrestricted use of their card, which allows them subsidized taxi rides; residents of Barnet or Redbridge, where the local Council operates an independent scheme, or of Haringey and Islington, where spending restrictions have hit hard, are restricted to a few trips per year on their Taxicard.

Choice in transport is still to a large extent determined by *income*. Income determines the level of privacy and comfort that one can purchase. As the car economy grows out of control, this is now the norm. The old image of the 'City gent' in his bowler, who might have been a manager or a clerk, emerging from the 8.10 from Surbiton, is outdated. A manager in the 1990s will have a company car, subsidized by the taxpayer. The two-car suburban or 'yuppie' household experiences transport in a quite different way from a pensioner, student, part-time worker or unemployed person who may have little transport 'choice': bus, the proverbial Tebbitian bicycle, or walking. Cost is a major consideration in choosing whether or not to travel: rail services, for example, are prohibitive for many low-income residents. Poorer people are put at a double disadvantage: weekly or monthly tickets, especially those that include rail, Underground and bus, 'save' money – but only for those who can afford the initial outlay. Single tickets are more expensive; the poor pay more to be poor.

Women are the main users of public transport; they are also singularly ill served by transport services. In spite of claims of a 'post-feminist' 1990s, basic inequalities between women and men remain, affecting fundamental transport need. Women are poorer, are more likely to work part-time, and average female earnings are lower than those of men. Black and ethnic minority women remain largely in the lowest-paid jobs. Women are the majority of disabled people and elderly, mostly living on benefits, especially in Inner London. The domestic responsibilities of caring for children and running a household still fall largely on women; it is women who fit their hours of paid employment around the requirements of their family. In London, a large number of all households are headed by a woman. In the majority of two-parent families, traditional patterns have changed remarkably little: where men share in shopping, for example, it is likely to be by driving the car to the supermarket once a week, loading the shopping and driving home; the rest is done by the woman locally, as are other tasks of caring for a family. All this has to be 'fitted in' around available transport. In fact the form of transport used most frequently by women is walking, closely followed by the bus. Yet between 9.30 a.m. and 4.30 p.m., when many women juggling family and paid employment are waiting for their bus to jobs that give them no leeway if they are late, bus services evaporate. Services planned around the needs of commuters into the city cannot cater for the multitude of short-hop journeys that women make in going about their everyday lives: from home to nursery, to school, to work, and on the way home via the shops to school, the dentist, the nursery, often carrying small children or shopping on buses with high steps and no conductors.

Most women interviewed in the Greater London Council's *Women on the Move* survey (GLC Women's Committee 1985–7) quoted quite mundane problems in using public transport: the lack of storage space and the impossibility of getting a pushchair with a child in it onto a bus, lack of information and the poor design of British Rail and Underground stations – none of these design and access issues is beyond the ken of woman to solve, though so far it appears to have baffled the men who run London's transport. For women one issue looms larger than any other: the danger of travelling on most forms of public transport, owing to harassment, threats and actual violence from men which women experience when walking, travelling on buses, trains, the Underground and in taxis, especially, but not only at night. (For further discussion of this issue see Chapter 7.) Whilst transport operators blame society at large, perhaps merely reflecting a widespread male complacency about violence against women, to many women London's railways and Underground are no-go areas, at least at night. Some 25 per cent of women in London (the statistics differ little elsewhere) will not go out at all after dark for fear of attack, though women who work shifts or do night work – many of them black and migrant women – have no option. Many women remain determined not to be put under curfew by the threat of violence. One response to this state of affairs has been the setting up by local women's groups of women-only Women's Safe Transport schemes in a number of boroughs. Aimed at empowering women, not only by providing a safe means of travel, but by allowing women to work in a traditionally male sphere of employment, these schemes remain underfunded and undervalued. It is hardly surprising that the number of women car-owners should have increased substantially in recent years. The private car gives the flexibility, and to some extent the safety that buses and trains seem unable to provide. Paradoxically, it is those very people who least need a car, non-disabled males with no shopping or children to convey to and fro, who venerate the private motor car as a god and will not go anywhere without it.

With age comes loss of income and an increasing dependence for many people on public transport. London's pensioners benefit from free travel on bus and tube, and half-price travel on British Rail within the London area. Every year, pensioners' organizations have to fight for the retention of the free pass, which allows them to break down some of the isolation that elderly people cut off from their families experience in many western cities. The bus is the mode of transport most favoured by elderly people, many of whom have never been car drivers. Yet pensioners, also, are ill served by public transport. Few use the Underground or British Rail, except for special outings; limited off-peak services (again, the 9.30 a.m. 'Bermuda triangle' of buses) leave them at risk – Age Concern has argued that hypothermia caused by waiting at cold, unsheltered bus stops is a major cause of illness and death amongst elderly people. The fear of attack or accident, on buses and in the street, keeps elderly people at home – the number of injuries to elderly people reported at one North London hospital casualty department increased after one-person operation was introduced on local bus routes; few of the incidents which caused these injuries had been reported to London Transport. And transport is not free to all elderly people: those who rely on Dial-a-Ride have to pay the equivalent of a bus fare, though they would travel free if they could use an inaccessible bus; many pensioners have a Taxicard but rarely use it because they fear the cost.

There are 400,000 *disabled people* in London who cannot use public transport, according to a GLAD survey. Whilst other groups are discriminated against in their use of passenger transport, disabled people are effectively excluded. Disability is defined here not merely as a physical but a social condition: people are disabled by an inaccessible, ill-designed environment. The needs of disabled people are the last to be considered by transport planners, politicians, operators and campaigners alike. Segregated from the rest of the population, educated in separate schools, housed often in 'residential units', transported by ambulance, disabled people have only relatively recently begun to make political demands, including the right to transport. Many feel that direct action is the most effective means of political pressure and have joined in stopping Central London traffic with the Campaign for Accessible Transport.

Discrimination in transport provision is almost total: within London, many disabled people are unable to use buses, British Rail and the Underground. In some instances they are banned. London Underground's by-laws ban wheelchair users from travelling on all the deep-level sections of the Tube, supposedly because of potential problems if passengers had to be evacuated.

Disabled people who wish to travel by bus in London face a restricted choice: a few 'Mobility Bus' routes fitted with lifts, running once or twice a day on different routes each day; an hourly service between mainline stations; an accessible 'Airbus' to Heathrow; and two local routes in Ealing and Hounslow, both the result of imaginative thinking by local Councils. British Rail, in prioritizing the improvement of access to its lucrative Inter-City facilities, has only a smattering of 'accessible' local stations in the London area, with ramp access, or lifts that require a member of staff to operate them; there is no cohesive network. The options for independent travel are few. London Underground have at least begun to concede the principle of accessible stations on surface lines. But there is no funding.

The alternatives to public transport are limited: lifts from friends or volunteers, which place the disabled person in a relationship of dependency, or second-rate transport by local authority or health service ambulance. Disabled children, already disadvantaged by an inferior segregated education, waste hours every week in this way. At present the only real and dignified options are the door-to-door services, Dial-a-Ride and the Taxicard scheme, and the private car. Unfortunately, demand for Dial-a-Ride far outstrips supply; the London Dial-a-Ride Users Association estimated in 1988 that if every registered user of Dial-a-Ride were to ask for a lift, it would take 11 weeks to clear the backlog; what non-disabled person would wait 11 weeks for a bus? Taxicard, a system of subsidized taxi rides, is out of reach of the pocket of many disabled people surviving on benefits, and the number of trips each user is allowed is limited in many boroughs, usually to one or two per week. London lags far behind major European cities (Stockholm has an unrestricted Dial-a-Ride service, a majority of accessible stations, and a cheap Taxicard scheme; Paris's RER has a number of accessible stations). Hampered by unimaginative management, cumbersome technology (British-made lifts and wheelchair restraints are slow and awkward), an inability to listen to the disabled community (the London Transport Disabled Passengers Unit does not employ any disabled people), discriminatory by-laws, and an unwillingness to change, British Rail and London Transport limit themselves to a piecemeal, inconsistent approach: there is some doubt as to whether the extensions to the Jubilee and East London lines are to be inaccessible, the first because 'the existing section is not accessible', the second for the lack of £5m. and the political will.

Recent research has shown *race* to be a significant factor in determining people's use of and experience of transport (GLC Women's Committee 1985–7). A few examples will illustrate the issues. The black communities in London are, with the exception of a few areas, much more dispersed than they were. People still return to visit those areas where their communal needs can be met, for African or Caribbean food for example, for community activities, for facilities such as hairdressers, or to visit friends and families; shopping serves as social contact, especially for elderly people who speak little English. People will travel some distance across boroughs. Racist attacks against black and other communities are on the rise in London; attacks occur on people's homes and in the streets, but also on public transport – including attacks on staff. Information, however, is hard to obtain; the police normally do not issue race-specific statistics. Migrant workers in the catering and cleaning sectors in London, legal or illegal, working nights or early morning shifts, are least able to complain about reduced or unsafe services. Whilst London Transport have introduced network maps and information in German, French and Japanese for the benefit of tourists, they have shown no similar willingness to cater for London's many permanent residents whose first language is not English – there are no signs in Bengali at Bethnal Green station, though it borders an area that is the centre of the Bangladeshi community in the UK.

For women, elderly people, the black communities, and disabled people in London, transport is in a permanent state of crisis. Present crises in management, innovation and technology only add to the problems of a system that was never designed for the majority: only for a mobile, unencumbered male minority. Far-reaching, imaginative measures are required.

THE WAY FORWARD

A number of complete or partial blueprints for the future of London's transport have been put forward – by the local authority associations (London Planning Advisory Committee, London Boroughs Association, Association of London Authorities), by London-wide groups (Centre for Independent Transport Research in London, Friends of the Earth, London Boroughs Disability Resource Team) and by local user groups. There is some level of common agreement (outside government and the Department of Transport) on the basic elements of a way forward.

1. *Enhanced investment up to normal European standards.* Fares support is essential. Contribution through taxation is equitable; increasingly higher fares are not – nor are they effective. Any centrally funded investment and subsidy – on the grounds that the progress of the capital may benefit the whole country – must go hand in hand with local taxation. A tax on local employers (such as

operates in Paris, and was guardedly welcomed by no less than the Corporation of London (1990)) or a California-style local sales tax (*Economist* 8 December 1990, p. 51), would be the best options, in addition to the contribution that London residents have been paying for many years through local rates (now the 'Community Charge'). The contribution of areas on the edges of London, supplying the bulk of commuters, must be addressed. A Transport and Road Research Laboratory study now several years old calculated millions of £s of benefit would accrue annually if cheap fares were allied in London with bus priority and restraints on the private car (Oldfield 1977).

2. *A coherent infrastructure investment programme* which would embrace funding of the Central London Rail Study proposals, strong Channel Tunnel rail links, a co-ordinated bus-purchasing policy and the development of an accessible transport network.

3. *Unified transport and land-use planning* which would involve the revival of a London-wide authority embracing both transport and planning (and preferably the running of other essential services such as the fire services). A full democratization of the work of organizations such as the London Planning Advisory Committee would be a start. British Rail's London suburban services (ideally, unified with the London Underground) should come under this Authority's control.

4. *Stabilization*: bus services must not be deregulated. There should be an early decision on London's buses: whether to tender, to franchise or to democratically control; the writers would prefer the last.

In order to achieve these desired objectives, a lot more thought has to go into a whole range of specific areas. Detailed forward planning has not been a strong point with either operators or planners – or to some extent transport unions and passenger bodies. However, we may take heart from initiatives in a number of areas which together point to a more coherent future. The following is not a prioritized listing; progress in all areas is interdependent.

Buses to trams

Some years ago, the borough of Camden put forward a 'superbus' proposal for a bus priority route across Central London – a plan dropped for lack of political will by the then Greater London Council. Today, five West London boroughs are discussing similar notions for the Chiswick–Hammersmith–Kensington alignment. Semi-continuous bus lanes should be seen as the first phase in a process leading to a tram-like system (which might take the form of a tram, trolley or mixed-mode vehicle, all, of course, fully accessible). This alone offers the hope of relieving both the rail/tube and road congestion that is plaguing Inner and, increasingly, Outer London. Edwardian Inner London survives, and was all built around bus, rail and tram systems. London is unlikely to function physically without the re-introduction of something akin to a tram system – in reality an essential component in the transport infrastructure. The recent highly successful expansion of London's night bus network stands as a relevant precedent: a trunk network mirroring a future tram network.

Rail

There are moves in several localities towards the merging of British Rail and London Underground infrastructure, often involving light rail technology. It is proposed to extend the East London Line southwards onto British Rail track through Peckham, and northwards on a redundant viaduct (thoughtfully safeguarded by the local authority) to Dalston and possibly beyond. Croydon and Kingston both support plans for their Outer London centres to be linked by light rail services on British Rail tracks (in the case of Croydon, involving on-street running in some sections) (British Rail/London Regional Transport 1988). The Docklands Light Railway, in retrospect an underfunded, one-off system, may be extended south of the river to Lewisham. The Centre for Independent Transport Research in London has proposed a unified British Rail/London Underground network for London, embracing 26 new rail schemes (Centre for Independent Transport Research in London 1988).

An accessible network

A network of fully accessible stations and bus services is long overdue. It would provide a major boost to tourism. Bus services are easiest to alter, starting with local routes and key services into Central London. Disability organizations' demands for legislation making it obligatory for all buses purchased after 1992 to be fully accessible to all, including wheelchair users, may be furthered by proposed European legislation.

Whilst it is recognized that it is not possible to alter all of London's inaccessible rail and Underground stations overnight, a core network of stations (see Figure 4.1), refurbished with lifts or ramps and facilities for blind and partially sighted people, expanding on London Transport's own limited proposals for a 5–10-year programme of accessible core stations, would be a start. In addition, all new rail and tube lines, and extensions to existing lines, must be fully accessible. Integrated with an accessible bus network, and with taxis and local Dial-a-Ride services, they might form the first stage in the development of a fully accessible transport system that does not discriminate against disabled people.

Interchanges

Today's uninspiring interchanges (from Victoria or London Bridge to Finsbury Park) (London Regional Passengers Committee 1988) are essential to the functioning of London's transport; each borough should have its own transport interchange, offering easy change-over between train, bus and other modes (see Figure 4.1). Adequate and accessible passenger facilities, parking, drop-off points

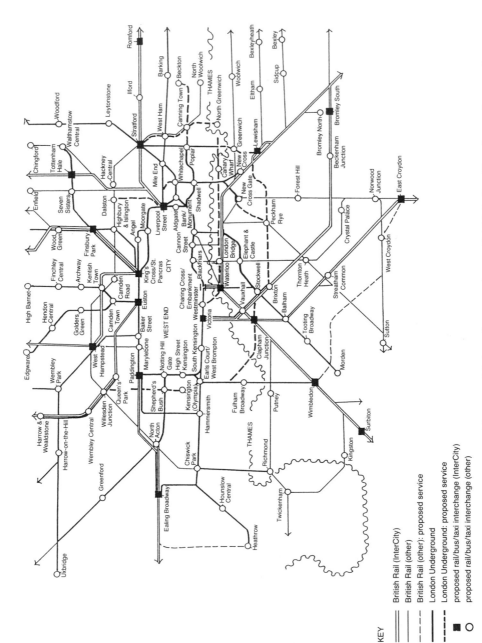

Figure 4.1 A proposal for accessible interchange stations on British Rail and London Underground

and taxi ranks should be provided by British Rail, London Transport and boroughs. This ties in with the 'core station' concept.

Staff

Restaffing of public transport is essential. An unstaffed, dehumanized system is an unsafe and unattractive system. Staffing levels must be increased not only on the Underground and British Rail so that no station is left without adequate staffing levels (and staff have adequate back-up when dealing with an incident, be it a suspected fire or a disturbance), but in essential areas such as track maintenance and cleaning, engineering, signalling and safety. Bus routes must, in Central London and on selected routes at least, return to safe and efficient two-person operation. The primary remit of many staff as mere revenue collectors must be called in question; the notion of staff as primarily passenger assistants, trained and remunerated accordingly, must be explored with some urgency. Overall, wages and working conditions must be improved if London's transport operators are to attract a stable and committed workforce.

Ticketing

A system of transferable bus–rail tickets that can be purchased, in book form for example, off vehicle is needed. A common practice in Europe, this would transform debates about ticket barriers and the rôle of staff. Travelcard is only a start.

Central London

We are in dire need of a European-style solution. This may not necessarily mean implementing a road-pricing system (a notion already being reconsidered by Stockholm), but must include stringent parking controls and an end to company-car subsidy. Wholesale bus (and future tram) priority and pedestrianization is essential, preferably on the Goteborg/Bremen model. (See Figure 4.2, based on Smith 1986 and 1987.) Two encouraging precedents are the Red Arrow buses servicing the rail termini, and the Corporation of London's tentative proposal to reduce road traffic at Bank. But current initiatives remain piecemeal; the importance of wider proposals with a strong image and their potential impact on London's decision-makers cannot be overstated.

Traffic calming

Dutch- or German-style traffic management is in evidence only as one-off schemes (Catford Broadway) or localized area schemes (Covent Garden, Soho, Poplar). In theory, the Department of Transport supports these as corollaries to their Red Routes, but progress in reality depends upon the boroughs (London Borough of Haringey 1990).

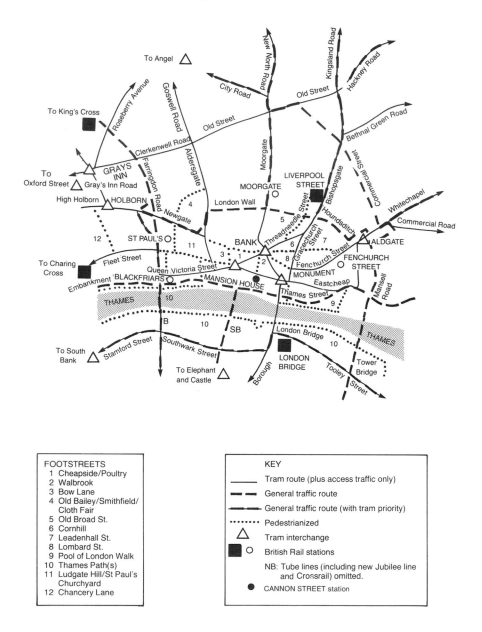

FOOTSTREETS
1 Cheapside/Poultry
2 Walbrook
3 Bow Lane
4 Old Bailey/Smithfield/
  Cloth Fair
5 Old Broad St.
6 Cornhill
7 Leadenhall St.
8 Lombard St.
9 Pool of London Walk
10 Thames Path(s)
11 Ludgate Hill/St Paul's
   Churchyard
12 Chancery Lane

KEY

——— Tram route (plus access traffic only)

– – – General traffic route

–·–·– General traffic route (with tram priority)

········ Pedestrianized

△ Tram interchange

■ ○ British Rail stations

NB: Tube lines (including new Jubilee line
    and Crossrail) omitted.

● CANNON STREET station

Figure 4.2 The City of London and its environs: suggested tram routes and footstreets

Comprehensive town centre schemes

Schemes involving bus priority, pedestrianization and traffic calming, and public transport interchanges, are beginning to be hinted at – though for the London boroughs, staff shortages in Planning and Engineers Departments act as a disincentive to this demanding work. As an example of their potential, we illustrate proposals from a borough-funded study of Central Hammersmith (Centre for Independent Transport Research in London 1991).

Walking

European-style pedestrian improvements, as at suburban centres like Harrow, Sutton, Peckham and Clapham Junction involving 'bus and access-only' streets, are slowly occurring. Environmental groups are an important lobby here (Planning Aid for London / Transport 2000, 1991).

Cycling

The government has accepted the aim of a 1,000-mile cycle route network, pushed by the London Cycling Campaign (1991) through the boroughs and all-Party parliamentary lobby. Children's safe routes to schools should be an important element in any cycling policy.

Demand-responsive services

Flexible transport services, many of them community- or locality-based, but all integrated in terms of funding into the main network: door-to-door services such as Women's Safe Transport or Dial-a-Ride, Taxicard, Community Transport.

Local democracy

A democratically elected London planning body is essential. The statutory passengers' watchdog – the London Regional Passengers Committee – should be retained, and keep its remit (contrary to the government's intentions) over buses as well as British Rail and the Underground. But we need local consumer committees too, to ensure that transport planning responds to need, not solely to a market-led notion of demand. The Lewisham Passengers Transport Users Committee is one example; others have involved, with varying degrees of success, the participation of local campaigns in liaison meetings with operators. The impact of these local initiatives can be only a limited one in the present undemocratic climate. Transport workers as well as users must be involved.

Freight plans

Local lorry bans in well-to-do suburban streets are an essentially negative approach to freight planning. Positive planning of cross-London lorry routes is required, and should be backed up by the development of trans-shipment depots, railheads and waterway freight. The night and weekend ban on lorries over 16.5 tonnes without a permit is a small beginning.

Legislation

The London Regional Transport Act will need to be rewritten to restabilize London's bus services. Other enabling legislation is required, most of it national: accessible transport, health and safety, including bringing London Underground under the aegis of the Railway Inspectorate; environmental assessment; and minimum service standards. A healthy precedent is the European Commission's threat to take the Department of Transport to court over its failure to assess the environmental impact of the proposal for a road carving through Oxleas Wood, a Site of Special Scientific Interest (*Evening Standard* 16 November 1990).

CONCLUSION

London's transport crisis is not an insoluble one. To arrive at a workable solution requires principally three things. The funds necessary to build and operate a modern transport system for a modern city must be made available. Professionals must make themselves aware of the notable advances evident in Europe. And all London's population and visitors must be given the right to move about in comfort and safety. The implied consumer rights charter can only come about if London's transport system is opened up to full democratic accountability.

NOTES

1   The figures refer to bus, tram and Underground operations only; other rail (and sometimes suburban bus) services are excluded.
2   See, for example, the Department of Transport's *Statement on Transport in London*, issued to boroughs in 1989.

REFERENCES

British Rail / London Regional Transport (1988) *Light Rail in London*, London, BR/LRT.
British Transport Police (1990) *Report*, London, BTP.
Bushell, C. (ed.) (1989) *Jane's Urban Transport Systems*, London, Jane's Information Group.
Centre for Independent Transport Research in London (CILT) (1988) *Railways for London*, London, CILT.
—— (1991), *Hammersmith Centre Traffic Study*, London, CILT.
Corporation of London (1990) *London's Transport: A Plan to Protect the Future*, London, Corporation of London.

Department of Transport / British Rail / London Regional Transport (1989) *Central London Rail Study*, London, DoT.

Department of Transport / Government Statistical Service (1990) *Transport Statistics for London*, London, DoT/GSS.

Fennell, D. (1988) *Investigation into the King's Cross Underground Fire*, London, HMSO.

GLC Women's Committee (1985–7) *Women on the Move*, London, GLC.

Hidden, A. (1989) *Investigation into the Clapham Junction Railway Accident*, London, HMSO.

London Boroughs Association (1990) *Capital Killer – Air Pollution from Road Vehicles*, London, LBA.

London Borough of Haringey (1990) *The Blueprint Route: Archway Road Corridor*, London, LB of Haringey. Produced in association with local community groups.

London Cycling Campaign (1991) *On Your Bike* (9th edition), London, LCC.

London Regional Passengers Committee (1988) *A Change for the Better*, London, LRPC.

Network SouthEast (1990) *Network Factfile*, London, British Rail.

Oldfield, R. H. (1977) *A Theoretical Model for Estimating the Effects of Fares, Traffic Restraint and Bus Priority in Central London*, Lab. Report 749, Crowthorne, Transport and Road Research Laboratory.

Olsberg, S. (1990) 'Gains for trains? Land value capture as a source of infrastructure funding', unpublished M. Phil. thesis, University College London.

Planning Aid for London / Transport 2000 (1991) *Traffic and Transport*, London, PAL/ Transport 2000.

Smith, G. (1986) 'Accessible city', *Architects' Journal* 9 April, pp. 32–8.

—— (1987) 'Civilising London's West End', *Architects' Journal* 24 June, pp. 34–48.

TEST / London Amenity and Transport Association (1984) *The Company Car Factor*, London, TEST.

5

LONDON AS ECOSYSTEM

Duncan McLaren

INTRODUCTION

Public concern about environmental issues increased significantly during the 1980s. Urban issues have not been high on that agenda, but concern has been expressed, particularly in London, over the increasing impact of traffic (see Chapter 4), while the desire to protect the surrounding countryside and to a lesser extent to 'green the city', by protecting open space, particularly 'wild' space, providing new spaces and cleaning up the street environment has obtained a higher profile.

These widely shared public concerns reflect a rather limited definition of environment, but even in this context London is in crisis: it has less open space than in 1970, areas of redevelopment such as the Docklands often provide less open space than the London average, the quality of many public open spaces has fallen as a result of cuts in local authority budgets, and sites of wildlife interest have been lost to development (McLaren 1989). Such a definition is, however, too limited. If we are to consider the environment of London in an adequately holistic manner, two concepts need to be introduced and briefly explained.

First, there is the concept of the urban ecosystem. This can be conceived as the human habitat – with which its inhabitants exist in some form of balance. But natural and semi-natural ecosystems are normally composed of largely closed cycles of resources, and are powered by the throughput of renewable solar energy which is transferred to different levels of the system as food energy. Urban systems, on the other hand (as extreme examples of human systems), involve one-way throughput of resources and rely to a great extent on the recovery of historical energy embodied in fossil fuels. Converting urban resource flows to cycles, reducing the disruption of global resource cycles and limiting energy dependence are basic principles of the second concept: that of sustainable urban development (Elkin *et al.* 1991). More generally this involves the application of general principles of sustainable development, as outlined by the World Commission on Environment and Development (WCED), to the urban system. Sustainable development is defined as that which

meets the needs of the present without compromising the abilities of future generations to meet their own needs ... [and] requires meeting the basic needs of all and extending to all the opportunity to fulfil their aspirations for a better life.

(WCED 1987: 8)

Although lip-service is paid to the basic concept by most politicians, very few have embraced it whole-heartedly. Sustainable development is clearly distinct from growth (even the concept of 'sustainable growth') in that it recognizes that production and income are not the only measures of human welfare. Even fewer politicians recognize that the concept of sustainable development should be applied to the city, despite the recent green paper on the urban environment from the Commission of the European Communities (CEC 1990) which called for detailed exploration of the implications of sustainable development for urban environmental management.

If we are to integrate the concepts of sustainable urban development and the urban ecosystem, we must understand what constitutes a 'successful' ecosystem. Although there is some debate, the *balance* of the system with climate and other external factors such as soil and geology is central. Natural ecosystems tend to achieve a developing balance with external factors which also provide constraints to the expansion of any particular ecosystem. The ability of a given ecosystem to support high numbers of individual creatures depends upon its *productivity*. However, there appears to be a trade-off between productivity and the ability of the system to survive external shocks (or its *stability*). Stability is also often related to the diversity, and therefore complexity, of the system. In the urban context economic diversity maintains economic stability, although often at the cost of sub-optimal productivity. Sherlock (1990) discusses the importance of diversity in the central areas of London, including the retention of residential functions. A stable but developing system is likely to be one in balance with its external environment and not in a situation of overproduction. This could also define sustainability.

It is important to examine ecosystems in terms of the relationships between species and their environment, rather than examining those elements in isolation. A key lesson of the urban ecosystem approach is the attention that must be given to how the city functions as a set of relationships between humans and their urban environment and between aspects of that environment such as its built form, transport systems or the urban 'green'. A good example of the integrated approach is provided by the work of the European Academy of the Urban Environment, Urban Ecology and Urban Open Space Planning (1989) which covers issues as diverse as food production, energy and waste. Sustainable urban development also requires an integrated approach to the city. Too often, narrowly defined political goals shift problems geographically or sectorally.

In summary, the traditional approach to urban 'greening' can be described as a concern for the ecosystem within the city, while we must also consider the city itself as an ecosystem and the city in the global ecosystem.

LONDON'S ENVIRONMENT

In discussing London's environment in this integrated way it is difficult to point to single clear indicators of crisis. But despite historical improvements in the environment experienced within the city, as a result of the Clean Air Acts or the improvement of sewage treatment, there is a crisis. The worst effects of environmental degradation become clear only in the long term, and the impacts of London on the wider environment have been increasing.

It is possible to apply ecosystem and sustainability concepts to London, but first we must briefly explore the nature and form of the contemporary urban system. Urbanism exists on a global scale, and individual urban systems are also of vast scale – London is not just 155,000 hectares of South-East England but is also and more importantly a socio-economic system which integrates physically and economically into a much larger area, much of which it dominates. Decisions made in London affect local environments everywhere from Westminster to Brazil and Antarctica, as well as impacting upon the global environment. The actual area covered by buildings simply constitutes the physical core of the urban system. This continuous and relatively densely developed core has expanded over time, although restrained more recently by green belts, but has also tended to become less dense, with a reorientation of many facilities and functions towards and beyond its fringes. Herington (1989) shows how this process has operated in the dispersal of London's population and other functions into the new towns and other south-eastern settlements which he described as the 'outer city'. Functionally, many of these settlements have become part of the urban system focused on London.

The rôle of the city as a crucible of political power has to a great extent been replaced by its key rôle in economic power, although in political centres such as London the political and economic power reinforce one another. However, the competition to maintain economic power can lead to environmental damage. Harvey (1990a) demonstrates the extent to which cities increasingly compete in an international arena for the investment of mobile capital in their locality. This is not only likely to generate extra transport movements but has serious implications for the discretionary environmental standards imposed on new developments through the planning system. In Urban Development Areas such as the Docklands, it has been argued that the transfer of planning powers has lowered standards, but local authorities challenged by developers prepared to go elsewhere have also relaxed standards. For example, Havering Borough Council in East London granted planning permission in 1990 for MCA's proposed film studios and theme park development at Rainham Marshes, despite the fact that the marshes are London's largest Site of Special Scientific Interest and that the nature conservation value of much of the site would be destroyed by such a development. If it were to go ahead, the development could also be expected to generate vast amounts of additional traffic (with associated pollution and congestion).

LONDON'S URBAN ECOSYSTEM

In this section we examine the environment experienced by the 6.8 million people living within the area of Greater London. The quality of this environment and the extent of its degradation will be examined with reference to the home and street environment, air and water pollution, land contamination and dereliction, development pressures and the threat to green space and wildlife.

In 1986, when the English House Condition survey was last carried out, it recorded for the first time the proportion of houses in poor environments (Department of the Environment 1988). A dwelling was considered to be located in a poor environment if action was required on four or more of the following concerns: roads or pavements in disrepair; fences, gardens or common areas in neglect; poor lighting; lack of trees or landscaping; presence of vacant or derelict sites; lack of public green space, or of private gardens; or intrusive industry or traffic. In regional disaggregation, Inner London proved to contain the highest proportion of such dwellings, over one-quarter of the total stock. In Outer London, the respective figure, at 13.3 per cent, was higher than the national average of 10.6 per cent. In Greater London as a whole, of the total stock of 2.8 million, 520,000 dwellings fall into this category (Department of the Environment 1988). The actual quality of housing is also a key element of the quality of the environment experienced by all people.

Many of the factors considered in assessing local environmental quality relate to the street environment. Amongst these are the effects of increasing volumes of traffic on many streets, which impacts both on through routes where severance for pedestrians is most extreme, and residential streets used by 'rat-runners' avoiding congested through routes. Traffic-calming is a key element in improving street environments through discouraging and slowing traffic, encouraging cycling and walking and releasing space for environmental improvement (Hass-Klau 1990). (These issues are addressed in more detail in the last chapter.) Streets are also degraded by air and noise pollution, increasing volumes of litter, dog-fouling, vandalism and the simple lack of greenery. Litter problems are most severe in Central London, where 10 per cent of the refuse collected in the City of Westminster is taken direct from the street, peaking at 90 tons per day in summer (Hillman 1988). Streets and informal open spaces are also the destination of the estimated million tonnes of illegally tipped (mainly construction) waste in London (London Research Centre 1988).

The quality of air in London is difficult to assess accurately, as statistics are aggregated for publication and monitoring stations are limited in number and often inappropriately located. For example, the Central London monitoring station for nitrogen dioxide, rather than being located in a street with traffic, and therefore measuring the levels to which pedestrians are actually exposed, is located in an alley where only background levels can be measured. A survey carried out by Friends of the Earth in 1989 indicated that European Commission 'guide' and 'limit' values may actually be exceeded in many parts of London.

(The lower guide values are advised levels of exposure, while limit values are legal limits established by the Commission directive.) Figures from one month's monitoring revealed average levels of 115.9 micrograms per m³ on High Holborn and 100.7 micrograms per m³ outside the Royal Free Hospital in Camden, compared with the annual average limit value of 84 micrograms per m³. London Scientific Services surveys, reported by the London Research Centre (1988) show that in addition to the European Commission nitrogen dioxide limit, World Health Organization guidelines for carbon monoxide and GLC guidelines for ozone have been regularly exceeded.

Large urban areas such as London have distinctive local climatic effects which can, in turn, significantly modify concentrations of pollutants. The key effect is the development of the urban heat island, due to greater absorption of incoming radiation, and the escape of waste artificial heat energy (Goudie 1981). In extreme conditions of night-time temperature inversion, air flows into the city from all directions. Pollution from peripheral industry can then be trapped and concentrated in the city by the inversion.

The effects of the urban surface on winds are complex. Chandler (1965) found increased wind-speeds in Inner London when prevailing winds were weak; but when prevailing winds were strong, wind-speeds in the city were lower than outside. Localized high-wind effects cause inconvenience in 'canyon' streets, but the general lower speed of winds means that local pollutants are less rapidly dispersed. Such urban pollutants include ground-level ozone produced by photo-chemical reactions, and this can be particularly problematic in summer. Such problems require stringent controls, particularly on the numbers and emissions of motor vehicles, but can be eased by provision of green belts and wedges which help absorb particulate pollutants and lower temperatures (Brown and Jacobsen 1987).

The heat-island effect tends to actually increase energy use in cities. Although figures for London are unavailable, a study of twelve US cities found that while heating was required in the city centres on 8 per cent fewer days than in the outskirts, the more energy-intensive process of air-conditioning was required on 12 per cent more days (Brown and Jacobsen 1987).

Urban pollution is not limited to that in air. The Robens Institute (reported by London Research Centre 1988) found that about 80 per cent of London's drinking water failed to meet European Commission standards, owing to excessive levels of lead, aluminium, manganese, other metals, and nitrates. A national survey by Friends of the Earth revealed 298 occurrences where single pesticides breached European Commission Maximum Admissible Concentrations under the Drinking Water Directive, of which 63 were in the Thames Water Authority area (Lees and McVeigh 1988). Yet the Thames Water Authority and its successor Water Supply Company were granted 'derogations' which allowed them to breach European Commission standards without risk of prosecution by the UK government. The worst excesses of river pollution in London are now past: it is over 40 years since the oxygen content of the Thames at Woolwich was officially zero (White 1984), but pollution is still a major problem in some tributaries of the Thames and in rivers in other urban areas.

Consistent and comprehensive surveys of the extent of land contamination have not yet been carried out in the UK. Estimates at the national scale suggest that a high proportion of derelict sites may be contaminated and that over 100,000 hectares of urban sites may be contaminated (Milne 1989). Among the range of previous uses which are likely to have left a legacy of contamination are waste disposal sites, petrol stations, power stations, chemical works, timber treatment works and gasworks. All of these can be found in London on sites ranging from Marsham Street (home of the Department of the Environment) to Beckton in Docklands.

Estimates of derelict, vacant and underused land vary widely with the definition employed and the concerns of the estimators. Official figures from the derelict land surveys of 1982 and 1988 suggest that dereliction in London fell to 1,386 hectares in 1988 compared with 1,954 in 1982 but only 324 hectares in 1974 (Department of the Environment 1989). However, it is widely accepted that the tight definition used officially ignores much waste and underused land. At a national scale 'derelict' land is thought to comprise approximately one-third of the wasted land resource (Civic Trust 1988). Other estimates suggest that this holds true in London: the London Research Centre (LRC) reports that in 1981 the Greater London Council (GLC) identified 4,720 hectares of unused and damaged land (compared with the figure of 1,954 from the Derelict Land Survey of the following year). In 1987 there were 1,896 hectares of underused land in London on the Public Land Register (Chisholm and Kivell 1987), but this excludes all privately owned underused land and public land which the holders wish to keep from the register. The LRC estimates that there may in fact be 8,800 hectares of unused or derelict land in London (1988). A proportion of derelict land is of nature conservation value, and several derelict sites fall into the group of Sites of Metropolitan Importance for Nature Conservation lost to development; but, in general, such derelict land should be developed, or at least converted into a public amenity. Of related concern is the wasted resource of vacant housing, which reaches 7.6 per cent of the stock in Inner London. While the Outer London figure of 3.6 per cent is below the national average, the overall result is 146,000 empty dwellings in Greater London (Department of the Environment 1988).

The basic consequences of these wasted resources include increased demand for development land, and therefore pressure on previously undeveloped green space of amenity of nature conservation value, and generally increased use of energy and resources in low-density urban development (Newman and Kenworthy 1989) plus increased demand for building materials. Furthermore, the speed of redevelopment of the urban fabric is generally increasing: many new buildings have an expected life-span of no more than 15 years. Harvey (1990b) demonstrates how this may be an expression of the accumulation dynamic of investment capital. (For a more detailed discussion of the resource and environmental impacts of built development see Elkin *et al.* 1991.) Development pressure is placed on sites at the urban fringe and within the city. Government commitment to green-belt protection, although more symbolic than significant (Friends of the Earth 1990) has led to development being deflected beyond the green belt (Herington 1989) but also to increased pressure on open space in highly developed areas. Figures are not easily available, but concern over the loss of playing fields and public open space is growing. Even the government has been convinced of the threat. In October 1989 the Secretary of State said:

> our policy for the use of urban land [for housing] is not a call to sacrifice the playing fields, allotments and private gardens which provide such valuable opportunities for recreation in our towns and suburbs.
>
> (Patten 1989: 5)

It is not, however, contradictory to argue for adequate green-space provision and high-density cities: nor does the resolution of that apparent conflict require high-rise living. Sherlock (1990) shows how urban population densities of 375 per hectare (50 per cent above those of much contemporary development) can easily be achieved with medium-rise development. Yet because of the high total demand and the inappropriate form of much development, open space is under severe pressure. Nine sites of conservation importance were lost in London in 32 months (1984–7) and twenty-nine were under threat at the end of that period. Sixty-two sites of amenity value were lost or threatened (McLaren 1989).

Green space is a key element of the diversity of the urban ecosystem, as it is often in very short supply. Around 11 per cent of London is open space, although this falls to 6 per cent in the Docklands area. (This compares with 17 per cent dedicated to transport uses.) We must, however, be cautious in interpreting such data. There is a wide literature that seeks to 'explain' economic success by the attraction of 'pleasant environments' for entrepreneurs (Fothergill and Gudgin 1982) or to explain levels of crime as a result of particular architectural designs (Coleman 1985). However, the availability and quality of green space are only one element of the urban system. To assert that these factors can explain (or to claim that they cause) such effects is gross environmental determinism and must be avoided. We cannot divorce economic activity from the action of the international economy and, in particular the multinational companies which influence it. Nor can we disregard the social circumstances and political decisions which influence crime levels. Having made this caveat, it is important to recognize that, in comparison with Amsterdam, for example, much of London lacks high-quality open spaces. In Amsterdam, many residential streets are traffic-calmed, while, for example, the Bjilmermeer estate – largely high-rise – is set in naturalistic landscaping, including woodland and wetland.

London's green belt cannot compensate for its lack of local green space. Moreover, green belts provide recreational opportunities only for particular groups of people. In 1973 the Countryside Commission surveyed visitors to country parks in the London Green Belt: none of them came from deprived inner boroughs (reported by Nicholson-Lord 1987). Leisure time and travel opportunities for deprived groups have improved since 1973, but not significantly. To a great extent, access to out-of-centre recreational facilities has depended on access to a car. In Britain as a whole in 1987, 37 per cent of households (46 per cent in 1973) and a much higher proportion of individuals had no access to a car, but of households headed by unskilled manual workers 62 per cent had no car and in London 41 per cent of all households had no car. Meanwhile, survey work by Harrison and Burgess (1988) in Inner London revealed that all groups, regardless of social class, income or place of residence, and including women and minority groups, valued local parks highly and viewed green space as an integral part of the urban environment. Yet the creation of new green space in development is limited. Developers have begun to recognize that appearing to be 'green' can increase profits, but on the ground such a motive has largely resulted in purely cosmetic improvements. Mabey (1989: 1) speaks scathingly of the Docklands approach, which, he claims, 'has made barely a single gesture towards the kind of green space now regarded as a prerequisite of tolerable urban life everywhere else in the world'.

Effective linking of green belt areas with green space within the city is needed to achieve the best results in terms of wildlife protection, providing corridors for movement, and in terms of amelioration of local climatic and pollution effects. Strategic planning is required to achieve this, using and developing the concepts of protected Metropolitan Open Land and 'green chains'.

LONDON IN THE WIDER ECOSYSTEM: DISRUPTION AND DISCONNECTION

The preceding section described the extent to which the environment experienced daily by Londoners has been degraded. But in environmental terms we cannot discuss London in isolation – the urban system of London is a key part of what Nicholson-Lord (1987: 212) describes as the planetary city: 'its factories in the third world, its banks in the west, its parks in the rain forest or the ice-cap'. In this section we will examine more fundamental aspects of the resource demands generated by London: and the crisis of unsustainability. The introduction to this chapter discussed the urban ecosystem concept. There we saw that the balance achieved between ecosystem and 'environment' is central to any measure of ecosystem success. We now examine the extent to which such a balance has been and could be achieved.

Urban areas affect the hydrology of the region through water abstraction to supply urban needs which may reduce river flows, thus affecting wildlife; and through the increased impermeable area. The latter leads to more rapid run-off and thus increases the risk of minor flooding, reduces recharge of natural watercourses in urban areas, and increases the proportion of their flow that is composed of urban effluent and the pollution of water by run-off from hard surfaces contaminated by oil and other pollutants (Goudie 1981).

Problems of litter and flytipping in London form only the tip of the waste iceberg. Natural and semi-natural ecosystems are characterized by the internal cycling of nutrient resources through the different parts of the system. The concept of waste is virtually alien to the system as a whole. On the other hand urban systems involve large amounts of waste. In the London boroughs the tonnage of domestic waste generated per capita is higher than for any other type of local authority in the UK (Chartered Institute of Public Finance Accountancy 1988). London is running short of space to landfill its waste; the LRC (1988) estimates that only 10–15 years' reserve space exists. It is also widely acknowledged that historic standards of environmental protection in landfill are inadequate (Elkin *et al.* 1991). In comparative terms a very low proportion of London's waste is recycled, even though the recyclable elements include paper, glass, metal, some plastic, and food and garden waste and constitute up to 60 per cent of the domestic waste stream. The total proportion of domestic waste recycled in London is highest in the Borough of Richmond, at around 8 per cent, which compares well with a UK average of just 2–3 per cent. But on an international scale the comparison is poor. Portland, Oregon, achieved 22 per cent in 1986, while in the Netherlands as a whole, over 50 per cent of aluminium, paper and glass are recovered (Elkin *et al.* 1991).

To achieve such a level of recycling the markets for secondary materials need to be stimulated, while separated collection of waste, as implemented in many Canadian cities (Brown and Jacobsen 1987), is required, rather than recycling deposit facilities. When the full costs of current treatment and disposal methods such as incineration or landfill are accounted, such policies to encourage recycling are likely to prove highly cost-effective as well as environmentally beneficial (Elkin *et al.* 1991).

The resources wasted include the nutrient value of much domestic waste and of sewage sludge. Nationally around 40 per cent of sludge is 'disposed of' to farmland, where much of its nutrient value is recovered. Very little of London's sludge takes this route, owing to its contamination with heavy metals from industrial effluents, its concentration in a few locations and the general lack of treatment facilities – such as large-scale composting plants – which would convert it into a transportable fertilizer. The UK's response to the European Commission decision to halt dumping of sewage sludge in the North Sea has been to propose incineration of sludge, with the air pollution risks that entails. Separation of industrial contaminants from sewage would allow composting of sludge to provide a relatively odourless, transportable and storable fertilizer. Heidelberg composts over 30 per cent of solid waste through municipal and individual composting, thus recovering its nutrient value; while in Sweden a dual sewerage system allows easy treatment of sewage for use as fertilizer.

The local climatic effects of cities have been discussed above, but cities also contribute to climatic change at international and global scales. Global warming is now acknowledged as a major threat. This is a general increase in temperatures caused by greater retention of incoming radiation as a result of increased atmospheric concentrations of certain gases – i.e. an enhanced greenhouse effect (Inter-governmental Panel on Climate Change 1990). Carbon dioxide is the major contributor, responsible for 55 per cent of the effect. Methane, nitrous oxide, chlorofluorocarbons (CFCs) and hydrofluorocarbons (HCFCs), while all having greater warming effects per unit, are emitted in much lower volumes. The rates of warming expected to result from continued increase in emissions (or even stabil-ization of emissions) are in excess of any experienced in global history. Previous climatic changes have resulted in major ecosystem migration. However, existing infrastructure fixes the location of urban ecosystems, and indeed, the framework of urban and agricultural land uses will greatly hinder the future geographical migration of natural or semi-natural ecosystems in response to climatic change. To achieve a balance between urban systems and their environment requires significant cuts in emission of CO_2 and other greenhouse gases. The IPCC suggests that a cut of over 60 per cent in CO_2 emissions is needed, simply to stabi-lize atmospheric concentration. Energy use in buildings and urban transport accounts for a high proportion of the total of 554 million tonnes of CO_2 output from delivered energy in the UK. Agriculture, on the other hand, accounts for just 0.9 per cent of the nation's energy consumption. Cuts in emissions must be achieved through the urban uses of energy. Analysis carried out for Friends of the Earth indicates that (in the non-transport sectors) measures designed to improve the efficiency of energy use and supply are relatively cost-effective (Jackson and Roberts 1989).

Energy use in space heating is a key concern. Jackson and Roberts estimate that space heating alone is responsible for over 25 per cent of the UK's carbon dioxide emissions. London's housing stock is ageing and thus very inefficient. Of Inner London's dwellings 49 per cent were constructed before 1919; 68 per cent of Outer London's were constructed before 1944; and over 80 per cent of the total predate 1965, when energy efficiency standards were first introduced (Department of the Environment 1988). Clearly only a very small proportion of dwellings will meet current energy efficiency standards, while even those stand-ards are only equivalent to those in force in Sweden in 1935. In the UK as a whole, many dwellings lack even basic energy conservation measures, although they could be applied. In 1982 60 per cent lacked draught stripping of any doors and windows; in 1986, 50 per cent had no, or less than 2 inches of, loft insul-ation, 66 per cent lacked double-glazing and 86 per cent lacked cavity-wall insulation (Hillman 1984; Department of the Environment 1988). Yet for an annual cost less than the amount that would be raised by placing VAT on domestic fuel, a twenty-year programme of refurbishment could improve the homes of everyone in fuel poverty, simultaneously improving their welfare and reducing their contribution to global warming (Boardman 1990; Johnson et al. 1990).

The development of alternative forms of energy supply also offers reductions in emissions. District heating, using waste industrial heat or based on combined heat and power, is widely used in Europe – examples include Malmo and

Helsinki, where significant reductions in emissions of sulphur dioxide and nitrogen dioxide resulted. However, some of these systems are based on burning domestic waste, and this requires the retention of recyclable materials in the waste stream, notably paper, and also risks emissions of various products of incomplete combustion, some of them carcinogenic. The major waste incinerator in London, at Edmonton, currently fails to meet European Commission standards for large combustion plants. Less risky technologies are also available. The results of a demonstration project in Berlin using photo-voltaic panels to utilize solar energy indicate that up to 30 per cent of the city's electricity could be generated in this way.

In the transport sector, modal shift away from private motor vehicles to mass transit, walking and cycling can produce far greater reductions in emissions than increased engine efficiency (Elkin *et al.* 1991; Robertson 1989). In the longer term major reductions in emissions from the transport sector can be achieved through planning policies designed to reduce the need to travel and to encourage higher densities of development. Sherlock (1990) outlines the case for higher-density cities which can support local facilities and employment opportunities. Newman and Kenworthy (1989) have analysed the factors influencing energy use in transport in large cities. They classify London as a city of low automobile dependence, although this is more accurate in terms of comparison with American and Australian, than with European cities. Also their analysis is based purely on the area of the Greater London Council and thus ignores much of what might be described as the outer metropolitan area. In Amsterdam, where central-area population densities are higher, car ownership is higher than in London but per capita fuel use is one-quarter lower.

This scale of reductions in private car use would provide the benefits of improved street environments as well as helping to reduce the probable impacts of climatic change. But it requires strategic planning and co-ordination of land-use and transport development in London.

CONCLUSIONS

The environmental crisis is not limited to London, it is not necessarily most visible in London, nor can it be solved purely by London. The crisis reflects wider trends in national and global social and economic development, and without change in the nature and form of urban development it cannot be tackled. Policies exist which can be implemented at all political scales: local authority, city-wide and national government, some of which have been outlined above. Elkin *et al.* (1991) set out in some detail policies which would promote sustainable urban development. Central among these are planning policies which will encourage the development of a city which is not only less wasteful of energy resources in buildings and transport, but provides a more livable environment for its inhabitants. The European Commission (CEC 1990) offers a similar analysis, suggesting that the root of environmental degradation of cities lies in functional separation –

and the travel needs which that generates – and in the internationalization of the economy, with resulting inter-city competition.

For London therefore, the development path must be to increase its diversity and density, and to attempt to revive its local economies, particularly around focal points in mass transit systems. Urban regeneration should focus upon improving housing, particularly in terms of energy efficiency, and improving the street environment through traffic calming. Redevelopment should aim to produce flexible long-life commercial buildings around decentralized focal points, particularly using derelict land (after effective decontamination). These policies reduce pollution through reducing use of cars and improving energy efficiency, but must be complemented by strategies to control other pollutants and for waste recycling. Throughout London and on the fringes of the built-up area, co-ordinated development of the open-space and wildlife resources could produce more accessible and higher-quality open spaces. Without such policies London will become even less desirable a place to live and unsustainable dispersal will continue. The resources already invested in London's buildings and infrastructure will largely to go waste and its contribution to global warming will continue to rise.

To conclude, the future of London must be people-orientated; to return to the definition of sustainable development, we should aim to meet basic needs and provide opportunities for all to fulfil their aspirations. London should meet the needs of all its inhabitants for employment, entertainment, education and a high quality of life, rather than being structured to the needs of international capital investment. This can only be achieved with participation by the people in the decisions and developments that effect their lives.

REFERENCES

Boardman, B. (1990) *Fuel Poverty and the Greenhouse Effect*, London, Glasgow and Newcastle, Friends of the Earth, National Right to Fuel Campaign, Heatwise Glasgow and Neighbourhood Energy Action.

Brown, L. and Jacobsen, J. (1987) *The Future of Urbanisation*, Washington, Worldwatch (Paper 77).

Chandler, T. (1965) *The Climate of London*, London, Hutchinson.

Chartered Institute of Public Finance Accountancy (1988) *Local Government Trends 1988*, London, CIPFA.

Chisholm, M. and Kivell, P. (1987) *Inner City Waste Land*, London, Institute of Economic Affairs.

Civic Trust (1988) *Urban Wasteland Now*, London, Civic Trust.

Coleman, A. (1985) *Utopia on Trial*, London, Shipman.

Commission of the European Community (CEC) (1990) *Green Paper on the Urban Environment*, EUR 12902 EN, Brussels, CEC.

Department of the Environment (1988) *English House Condition Survey 1986*, London, HMSO.

—— (1989) *Review of Derelict Land Policy*, London, Department of the Environment.

European Academy of the Urban Environment, Urban Ecology and Urban Open Space Planning (1989) *Discussion Papers*, Berlin, EAUE.

Elkin, T., McLaren, D. and Hillman, M. (1991) *Reviving the City*, London, Friends of the Earth.

Fothergill, S. and Gudgin, G. (1982) *Unequal Growth*, London, Heinemann.

Friends of the Earth (1990) *How Green is Britain?*, London, Friends of the Earth.

Goudie, A. (1981) *The Human Impact*, Oxford, Blackwell.

Harrison, C. and Burgess, J. (1988) 'Qualitative research and open space policy', *The Planner* 74, No. 11.

Harvey, D. (1990a) *The Urban Experience*, Oxford, Blackwell.

—— (1990b) *The Condition of Post-Modernity*, Oxford, Blackwell.

Hass-Klau, C. (1990) *An Illustrated Guide to Traffic Calming*, London, Friends of the Earth.

Herington, J. (1989) *The Outer City*, London, Harper & Row.

Hillman, M. (1984) *Conservation's Contribution to UK Self-sufficiency*, PSI and RIIA Joint Energy Programme Paper No. 13, Aldershot, Gower.

—— (1988) *A New Look for London*, London, HMSO.

Intergovernmental Panel on Climate Change (IPCC) (1990) *Policymakers' Summary of the Scientific Assessment of Climate Change*, WMO and UNEP, Geneva.

Jackson, T. and Roberts, S. (1989) *Getting Out of the Greenhouse*, London, Friends of the Earth.

Johnson, P., MacKay, S. and Smith, S. (1990) *The Distributional Consequences of Environmental Taxes*, London, Institute of Fiscal Studies.

Lees, A. and McVeigh, K. (1988) *An Investigation of Pesticide Pollution in Drinking Water in England and Wales*, London, Friends of the Earth.

London Research Centre (1988) *London in Need*, London, London Research Centre.

Mabey, R. (1989) in *The Sunday Times Review* 12 February.

McLaren, D. (1989) *Action for People?*, London, Friends of the Earth.

Milne, R. (1989) 'Contaminated land poses new problem', *Planning*, 23 June.

Newman, P. and Kenworthy, J. (1989) *Cities and Automobile Dependence*, Aldershot, Gower.

Nicholson-Lord, D. (1987) *The Greening of the Cities*, London, Routledge & Kegan Paul.

Patten, C. (1989) *Planning and Local Choice*, London, Department of the Environment.

Robertson, T. (1989) 'Changing transport policy to combat air pollution from vehicles', *Proceedings of seminar F of the PTRC European Planning and Transport 17th Summer Annual Meeting*, London, PTRC Education and Research Services.

Sherlock, H. (1990) *Cities are Good for Us*, London, Transport 2000.

White, J. (1984) *Country London*, London, Routledge & Kegan Paul.

World Commission on Environment and Development (1987) *Our Common Future*, Oxford, Oxford University Press.

6

CLASSIC CARBUNCLES AND MEAN STREETS

Contemporary urban design and architecture in Central London

John Punter

INTRODUCTION: THE RECURRENT CRISIS OF ARCHITECTURE AND DESIGN

The debate over the quality of the built environment always intensifies in periods of rapid redevelopment, and the late 1980s in London were no exception. The general public, special interest groups, the media and built environment profes sionals and, most influentially, the Prince of Wales all entered the debate about the rapidity and scale of change in Central London, the loss of cherished town-scape and local communities, and the quality of the new architecture. The debate has many historic resonances, even with the post-fire debates of 1666 7 when the key opportunity for radical re-design of the capital was lost. But more recent parallels can be drawn with the protests that culminated in the formation of the Society for the Protection of Ancient Buildings (1877), the Georgian Society (1937), the Victorian Society and Civic Trust (1957) and the Thirties Society (1979). There are exact parallels with 1973 when the publication of *Goodbye London* (Booker and Lycett Green 1973) identified 390 redevelopment projects that threatened the conservation of the city's historic character.

So if the concept of a crisis of architecture and urban design in London is a recurrent phenomenon, what factors distinguish this particular crisis in the built environment from those of previous years? What have been the key foci of the debate about urban design and architecture and how have the issues been resolved if at all? And what is the current agenda for shaping London's urban environment? These questions will be addressed first by analysing the nature of the third post-war property boom, and the way that the development, design and planning processes have changed. Special consideration will be given to the inter-ventions of the Prince of Wales into the omnipresent battle of the styles. The emergence of privatized forms of urban design will be examined alongside the demise of government investment in the public realm and the sharp contrasts with current practice in European cities outlined. Finally, the whole question of

69

the strategic role of urban design and design control will be examined and its vital relationship to infrastructure explained as a prelude to examining what new responses are necessary to the perceived crisis of urbanism in London.

THE DEVELOPMENT AND DESIGN PROCESSES IN THE THIRD POST-WAR PROPERTY BOOM 1980-90

The third post-war development cycle began with a minor upturn in development activity as a response to the return of a Conservative government, but tight economic policy meant that it was not until 1984 that rising office rents encouraged a significant wave of development activity. In contrast with the investor-led 1970–3 boom, this time the upturn in development was user-led with new clients looking for a much more sophisticated product both in terms of interior space and exterior finish. The demand was met primarily by a new breed of development entrepreneurs (Figure 6.1) whose professionalism embraced extensive market research, innovative financing, new construction methods, new design techniques, a more enlightened approach to architectural quality and urban design, and a more innovative approach to circumventing the obstacles of planning and community protest (Rabeneck 1990a).

At the heart of this rising demand for office floorspace in Central London lay the growth of the financial services sector and the particular growth in international banking. The impending deregulation of the City set the seal on this growth as Japanese, American and European banks and finance companies awaited the opening of the UK Securities market to international traders in 1986, and the possibility of 24-hour trading using the Tokyo, London and New York stock exchanges.

These companies and their British counterparts demanded a new kind of office building with large 'footprints' (site coverage) to allow wide, clear, trading floors, with exceptional levels of servicing including telecommunications and computer cabling and air conditioning, thereby increasing storey heights (Duffy and Henney 1989). The technological sophistication required of these buildings was matched by the demand for quality interiors, private and semi-public amenities and, to some extent at least, quality architecture and urban design to the highest international standards. The large new floor plates required could rarely be shoehorned into the existing prime sites close to the Bank of England or the Stock Exchange and with electronic communications proximity was no longer an absolute necessity. So the developers turned their attention to the most accessible large clear sites where such buildings could be erected – the railway stations and sidings and the redundant Docklands (Figure 6.1). Then they turned to more complicated sites where building obsolescence offered significant opportunities – Fleet Street, Spitalfields, Smithfield – or the 1960s comprehensive redevelopment schemes like Paternoster Square and the London Wall.

The larger scale of the individual buildings was matched by a larger scale of overall project, many consisting of a sequence of buildings phased over several

70

years, frequently exceeding 100,000 square metres in size and often incorpor-
ating retail uses, improved transport interchanges, and a new public realm.
Some of these projects not only account for the most dramatically changing city-
scape of the 1980s, but the key development sites of the 1990s.

The new breed of developers utilized largely American fast-track steel
construction methods and increasingly used the larger American architectural
offices to cope with the scale and speed of design. So the globalization of international finance which drove the user demand thus led to a globalization of design
and a much resented 'American invasion' in design practice (Rabeneck 1990a).

CHANGES IN PLANNING AND DESIGN CONTROL: 1980–90

As the deregulation of the stock market drove demand so the deregulation of the
planning system, both nationally and within London itself, made larger scale and
more rapid development possible. The incoming Conservative government's
relaxation of planning controls began with Circular 22/80, which not only
emphasized that the 'onus of proof was on planning authorities to show why
developments could not be approved' but limited design intervention outside of
conservation areas. It invited aggrieved developers to appeal to the Secretary of
State, and throughout the 1980s both the number of appeals and their success
rate increased significantly (the latter peaked at 42 per cent nationally in 1988).

Specific elements of deregulation that directly affected London included the
designation of the London Docklands Development Corporation (LDDC) which
removed 'the greatest development opportunity in Europe' from local control
and placed it in the hands of a board appointed by central government. The
LDDC's grant of a planning permission for 400,000 square metres (now one
million square metres) of development on a 71-acre site at Canary Wharf in
1985, without a public inquiry, transformed all the planning rules for commercial
redevelopment in London in one way or another. Having significantly eroded the
negotiative power base of the planning authorities in the development control
process through circulars the government proceeded to abolish the Greater
London Council (announced in 1982, implemented in 1986), leaving London
without a genuine strategic planning authority. The London Planning Advisory
Group set up to advise the Secretary of State did develop a strategic view and has
begun to report on issues like high buildings policy, but their 1989 strategy was
rejected in favour of leaving London's development to market forces.

The net effect of all of these changes, and of the pressures generated by the
third property boom, were very clearly expressed in the redrafting of the City of
London Local Plan between 1984 and 1986. The Draft Plan had been ten years in
gestation and its conservationist tone was more of a response to the recession
provoked by the post-1973 property crash than to the impending pressures for
commercial redevelopment in the City. There would undoubtedly have been
changes in the plan given the strength of the opposition and the irrefutable logic

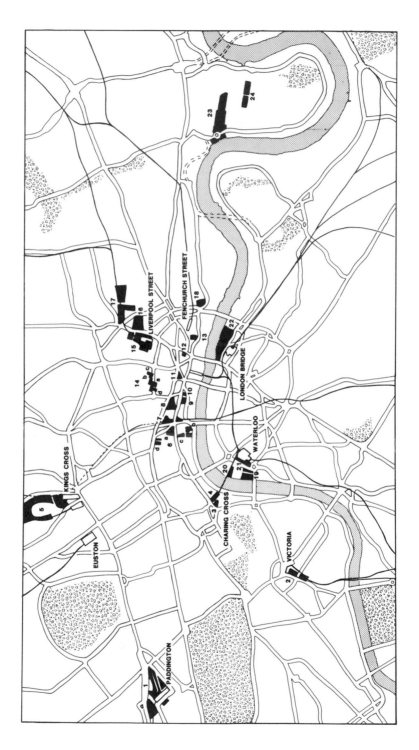

Figure 6.1 Major Central London commercial redevelopment schemes 1955–95

Name of development	Developer	Architects	Status	Approximate office space (millions of sq. ft)
1 Paddington Basin	Regalian, NFC, Higgs & Hill	BDP	P	0.67
2 Victoria Plaza	I Greycoat/British Rail	Arup	C	
	II Greycoat/McAlpine		C	
3 Grand Buildings	Land Securities	Siddell Gibson	UC	
4 Charing Cross	Greycoat/British Rail	Farrell	C	0.4
5 King's Cross	Stanhope/British Rail	Foster	P	5.87
6 Fleet Street				
a Daily Telegraph	Goldmans Sachs	Kohn Pedersen & Fox	UC	0.55
b City of London Boys' School	JP Morgan	BDP	UC	
c News of the World	Weatherall Green & Smith	YRM	UC	
d Daily Express	?	Fitzroy Robinson	P	
7 Farringdon/Holburn Viaduct, Ludgate Hill	Rosehaugh Stanhope	SOM, RHWL, Outram	C	0.6
8 Paternoster Square	Greycoat/Park Tower/Mitsubishi	Simpson	P	0.75
9 Petershill	MEPC	Cullinan	P	0.25
10 Bracken House	Financial Times	Hopkins	UC	
11 No. 1 Poultry	Peter Palumbo	Stirling	PP	1.3
12 Gracechurch Street	Barclay's	GMW	UC	0.45
13 Minster Court	Prudential	GMW	C	0.6
14 London Wall				
a Alban Gate	MEPC	Farrell	C	0.4
b Moor House	Greycoat		PP	0.4
c No. 1	Stanhope/Kajima		PP	
d Winchester House	Wates		PP	0.47
15 Broadgate	Rosehaugh Stanhope, now Stanhope	Arup (1–4), SOM (5–9)	C	4.5
16 Spitalfields	LWT/Balfour Beatty	Swanke, Hayden, Connell	P	0.9
17 Bishopsgate	LET/British Rail	Covell Wheatley Mathews, Hunt Thompson	P	1.3
18 Royal Mint	City Merchant/Postel	Sheppard Robson/RMJM	P	0.55
19 County Hall	New England/London & Metropolitan London Regeneration Consortium	SOM	PP	1.2
20 South Bank	South Bank Board/Stanhope	Farrell	P	0.3
21 Waterloo Street/York Road	P & O	?	P	1.2
22 London Bridge City	I & II St Martin's Property Company	Twigg Brown, J. Bonrington	C	1.0
	III St Martin's	Simpson	PP	1.25
23 Canary Wharf	Olympia & York	SOM, KPF, Pelli Pei	UC	10.0
24 South Quay/Thames Quay	Marples/Imry Merchant	Seifert/YRM	C	0.6

P = Planned, PP = Planning Permission, UC = Under Construction, C = Complete

Figure 6.1 (contd)

of their arguments that if the City was to compete as a world financial centre, then it would have to accommodate major financial services growth. However, the approval of the vast Canary Wharf Scheme on the Isle of Dogs is widely assumed to have panicked the City into a total pro-office development stance. The key changes made to the Local Plan in 1986 included the decision not to bargain for section 52 agreements for planning gain, the relaxation of the requirement for co-ordinated planning from scheme to scheme, the abandonment of special business areas restricting redevelopment, the withdrawal of the schedule of 'unlisted' properties of architectural merit that warranted protection, the creation of air rights for development over roads and railways, and the extension of the maximum 5:1 plot ratios across the entire City area (Marmot and Worthington 1987). (A plot ratio defines the allowable floor space as a ratio of the site area being developed.) These increased permissible densities were bolstered by exempting basements and not calculating gross floor areas in the plot ratio calculations. It was an invitation to large-scale redevelopment and the response was immediate and dramatic.

THE ARCHITECTURAL DEBATE: GLASS STUMPS AND CLASSIC CARBUNCLES

The architectural debate about the future of London was slow to grasp these new development and planning realities. It remained locked in the battle of styles that has been a feature of the British architectural psyche since the Victorian stylistic revivals. Architects had won a higher degree of design freedom in the wake of the Circular 22/80 and this very freedom became the target, between 1984 and 1990, of the Prince of Wales in a series of high-profile speeches, television programmes and a book. The twin foci of his initial criticisms were the proposals for a Mies van der Rohe twenty-storey tower opposite the Mansion House, which he referred to as 'yet another glass stump', and the winning entry for the National Gallery Extension by Ahrends, Burton and Koralek, which he described as 'a kind of vast municipal fire station ... a monstrous carbuncle on the face of a much loved and elegant friend' (Jencks 1988: 43). The Prince's command of the most accessible and cynical form of architectural criticism ('looks like ... reminds me of') apparently had the desired impact. The Mansion House Square, then at appeal, was rejected by the Secretary of State, while Sainsbury's, the sponsors of the National Gallery extension, withdrew their support from the scheme. The Prince certainly raised the profile of the architectural debate in Britain and especially London, but he reinforced the anti-modern prejudices of many people, and was responsible for largely re-confining the debate to issues of style and external appearance, often missing both the essential link between the development process and architectural form, and the fact that architectural quality is only one component of urban experience or urban quality.

But both the glass stump and carbuncle sagas had rather unexpected and deeply ironic endings. The former resulted in an important commission for James

Stirling but a major conservation battle that ended in the House of Lords with approval of the decision to demolish the eleven listed buildings in an 'outstanding' conservation area. The latter resulted in a rather tame piece of neo-classical infill by one of the high priests of pop art post-modernisms, Robert Venturi. While architectural critics revel in the contradictions inherent in this outcome, one cannot resist drawing comparisons with the National Gallery extension in Paris – I. M. Pei's bold, modern and innovative glass pyramid in the courtyard of the Louvre. These two schemes epitomize the relative confidence in modern architectural design in London and Paris.

THE CRUSADE FOR CLASSICISM AGAINST MODERNISM – PATERNOSTER SQUARE

The Prince continued his crusade against modernism and for classicism, and his campaign for tighter aesthetic control, in other key speeches in 1985, 1986 and 1987. Much of his 1987 speech revolved around the protection of St Paul's Cathedral as London's most potent symbol – national, religious, Wren-classical – and for the Prince especially a deeply personal touchstone. He concentrated his criticisms upon the proposals to redevelop Paternoster Square (Figure 6.2), one of three major tracts of the City to be comprehensively redeveloped after the war (Jencks 1988). Here the potential redeveloper Mountleigh (alias Stuart Lipton), seeking to reduce the planning difficulties and public controversies that always beset major London schemes, had organized a competition for redevelopment proposals. Seven élite architectural practices were invited to submit schemes and Arup's were selected to develop the masterplan for the site. When Lipton asked Prince Charles to comment upon the seven finalists, the Prince took issue with the brief for the site and advocated a return to the medieval street, low-rise buildings, 'and the ornament and detail of classical architecture' (Jencks 1988: 49).

In February 1988 the architect John Simpson, apparently sponsored by the *Evening Standard*, produced a classical scheme for the site that was a direct response to a brief developed by the Prince of Wales' architectural advisers. In June 1988 when Arup's unveiled their masterplan at a public exhibition their necessarily *evolving* designs for the majority landowner were overshadowed by Simpson's vast, precise and highly detailed model that appeared much more comprehensible to the general public, who were unaware that this scheme had no developer.

Whether it was the way they were being out-manoeuvred or the fact that they were given a surprisingly generous offer for the site, or both, that persuaded Lipton to sell may never be known, but the new developers (Greycoat – Park Tower – Mitsubishi) have announced their intention to use Simpson's ideas with their 1870s street pattern and full classical garb and have engaged Terry Farrell and Thomas Beeby to assist Simpson with the masterplanning. In an outline of the proposals on public exhibition in the summer of 1991 the six building blocks were designed by five different practices but all in derivations of the classical

Figure 6.2 View of the new Paternoster Square. 'The aim of the Masterplan for
Paternoster Square is to create a thriving new community in the heart of the City with a
variety of shops, restaurants and open spaces for the enjoyment of office workers and
visitors alike' (Paternoster Associates)

tradition and retaining Simpson's high site coverage and nineteenth-century
street plan.

Paternoster reveals many key issues about the content, form, style and process
of contemporary development in London. First of all there were criticisms of the
brief which proposed a 9.5:1 plot ratio, nearly twice that permissible in the City,
and nearly three times that of the existing post-war development. Secondly it
would appear that Simpson's reversion to the old street pattern and largely solid
building blocks provides the best solution to accommodating this quantity of
floorspace on the site. Thirdly there can be little doubt that the choice of fac-
simile classicism as a style did appeal more to the general public than Arup's more
modernist reinterpretations of the classical. As regards the process of design,
Lipton's attempts to generate a wider range of design solutions and encourage
public debate were widely applauded, but can also be seen as a way of obfu-
scating the issue of overdevelopment on the site. In the event the intention to find
a way of reducing the delays caused by public opposition was subverted by the
Prince's interventions, but Paternoster provides a valuable learning experience on
the pitfalls of more open design processes.

Meanwhile Simpson's success continued with a Venetian classicism scheme
replete with a St Mark's Square campanile for the third phase of London Bridge
City adjacent to Tower Bridge. Taking no chances in the 'battle of styles' the development company submitted three alternative designs to a public inquiry. The

inspector plumped for the Simpson scheme, ostensibly because of its urban grain and the spaces and streets that it created, though he confessed that he had some sympathy with the local MP's criticisms that the scheme would turn the area into an architectural theme park. By March 1991 John Simpson was threatening to resign as his designs for structural classicism were translated into 'clip-on' façades by another practice. This potential resignation is of course of great significance to the future of the Paternoster Square scheme and a key example of the drive for façadism and the commercial pressures on architectural quality in speculative development.

As in the 1920s, 1930s, 1960s and 1970s, when stripped classicism, neo Georgian, modernism and neo-vernacular respectively were promoted as the 'approved styles' by various élite groups, so the classicism strongly supported by the Prince of Wales is gaining ground in the 1990s. The debate about style is highly polarized, and not just in architectural circles; for many popular polls place key buildings like Richard Rogers' Lloyd's building in the lists of both the best and the worst buildings! What is evident is that there is a healthy pluralism in contemporary architecture, at least in the commissions given to the best architects. Their designs span the range of styles and approaches from high modernism and high-tech architecture through a variety of contextual-modern designs to the Beaux Arts and neo-classicists. The sheer diversity and richness of these forms of late-modern or post-modern architecture is greatly enlivening the townscape of London and stimulating public awareness of what is possible. Such pluralism is an essential prerequisite of a healthy urbanism. But whereas the work of these practices, and indeed to some extent the commissions themselves, reveal the positive side of the new development process, they tend to be beacons of quality in a sea of speculative-built mediocrity characterized by overdevelopment, banal elevations and bizarre profiles.

THE IMPORTANCE OF URBAN DESIGN: THE EXAMPLE OF BROADGATE

The problem with the preoccupation with architectural styles and elevations is that it has often prevented serious discussion of the broader urban design qualities of new development. It has focused attention on the visual imagery conjured by the architectural photographer rather than the user experience of the new townscape. Urban design is concerned with much broader issues of urban form (height, bulk, massing, etc.) but particularly with the public realm created by new buildings – the public spaces and streets, the semi-public spaces within buildings, ground floor activities, the furnishing, landscaping and micro-climate of spaces, that is to say with all the human senses not merely the visual.

One development that has seriously deployed the preoccupations of urban design is Broadgate (Figure 6.3). It stands as a testament to those who argue that the key to improved design in Britain is a recognition by developers that quality will ensure both higher rents and greater longevity for development. Stuart

Lipton's first dictum that 'good architecture makes money' has set new standards for architecture and urban design in London (note also his Stockley Park business space development north of Heathrow).

Broadgate is the showpiece London development of the 1980s, pioneering and epitomizing the new locational shifts, design and development processes that emerged in the second half of the decade. The first Rosehaugh Stanhope–Greycoat office building on the site designed by Arup's became the first speculatively built office building to win a major architectural award. Even more impressive in a speculative development was the acceptance of mixed uses (the offices have a restaurant, pub and shops on a public walkway through the scheme) and the creation of a square that is generous and calm, and furnished and landscaped to very high standards (Spring 1988). The first two phases of Broadgate completed the east and south sides of the square while the next three phases paralleled the western edge of Liverpool Street Station, creating a link with both the new station concourse and the Finsbury Avenue Square. A second new square doubles as an ice rink in winter and a multi-purpose performance space in the summer, fringed by a semi-circular arcade of shops and two restaurants which reinforce public use.

The quality of architecture of the first five phases of Broadgate does not match that of the first Finsbury Avenue buildings; but the treatment and continuity of

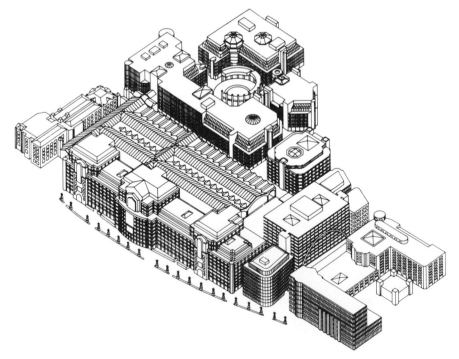

Figure 6.3 Broadgate development

spaces, the lavish furnishing and public art commissions, the active management in terms of cultural events, and high-grade maintenance are all exemplary. The later phases of Broadgate are severely compromised by an overdevelopment of the site consequent upon the rise in plot ratios allowed by the new City of London Local Plan in 1986. While the glass and cast iron train shed of Liverpool Street Station has been saved and extended, and a new square created at its northern end, plot ratios have risen to 7.5:1, meaning that the six to eight storeys of phases 1 to 5 have become twelve storeys on phases 6 to 8. The latter are treated as one huge megastructure towering over the traditional five storeys or less of Bishopsgate, Spitalfields and beyond. This vast megalith has been designed by the American practice of SOM and while it possesses an arcaded shop-lined walkway above the pavement, furnished and treated in best Beaux Arts tradition (ground-floor retail was a planning requirement), and a richness of elevations that help to disguise its size, its essentially pastiche nature and overblown scale make it a potent symbol of greed and arrogance to professional and lay critics alike (*Weekend Guardian* 25–6 February 1989). The smaller buildings beyond (phases 9 to 10) on Sun Street are only seven storeys and their warm terracotta colour, rich modelling and crisp fenestration complete a composition on Appold Street that is always interesting and intricate at a city scale (Rabeneck 1990b). Broadgate has its critics, but it has introduced North American qualities of urban design to office developments in the City, and it has set the standards by which future projects will be assessed.

URBAN DESIGN IN SPITALFIELDS

In the literal shadow of the Bishopsgate megalith of Broadgate stands Spital-fields, a development site for the mid or late 1990s, but one whose development history already poses important questions about planning, urban design and architecture in contemporary London. Spitalfields lies in the impoverished London Borough of Tower Hamlets, separated from Bishopsgate by a thin strip of land in the City of London. However, the vegetable market, the key development site, is actually owned by the Corporation of the City of London, compli-cating a development process already made difficult by market relocation, and by the juxtaposition of rich redevelopment potential and poor social conditions seldom equalled in the developed world. Spitalfields' development history illus-trates many important things. Late in 1990 the developers (Spitalfields Develop-ment Group) finally submitted their revised planning application called in by the Secretary of State because of the reservations of the Royal Fine Art Commission and English Heritage. Having successfully negotiated a land deal, development brief, Parliamentary Bill to move the market and market relocation, and survived strong competition from rival developers, long and complex negotiations on planning gain, the onset of the recession, as well as a whole host of design changes, the developers had every right to feel aggrieved.

However, the longevity, complexity and therefore cost of the negotiation process is not the main interest of the Spitalfields saga. From a design perspective it is the way architecture and architectural practices were used as pawns to try to legitimate the development, to disguise its overwhelmingly commercial imperatives, and even to attempt to subvert the leading scheme. For Spitalfields turned into a quite spectacular graveyard of competing design concepts as the lead developers attempted first to negotiate an acceptable design (Fitzroy Robinson joined by MacCormac Jamieson and Prichard), then to get a commercial design produced to a tight schedule (American architects Swanke Hayden and Connell), then to refine the scheme to meet community and conservationist interests (Cullinan, Hunt Thompson and Burrell Foley), and perhaps finally to please the most powerful critics (Farrell with Tibbalds *et al.*). The process led Richard MacCormac to coin the phrase 'Trojan Horse' architecture, whereby a scheme is designed by one practice to gain a permission and then redesigned by another to cut costs and increase commercial appeal (Tibbalds 1990).

But this galaxy of architectural practices was matched by those employed by Rosehaugh Stanhope, the developer of adjacent Broadgate, who attempted to subvert SDG's efforts with a rival scheme between 1986 and 1988. Rosehaugh Stanhope selected Leon Krier as their architect, even though he had never built any commercial buildings; perhaps he was chosen because he was known to be much admired by the Prince of Wales. Given Rosehaugh Stanhope's established working methods it was an extraordinary choice and prompted speculation that they were merely trying to delay the SDG scheme in order to ensure that the later phases of Broadgate, still being completed in 1991–2, did not have any local competition when they came on the market.

While Krier produced the masterplan, detailed designs were to be executed by the classicists Quinlan Terry and Robert Adam, and the more classically inspired post-modernists Jeremy Dixon and Terry Farrell (to emphasize the patronage/ design respectability game). Krier's low-rise scheme largely conformed to the brief and recreated the intricate morphology of the East End, but when asked to increase the quantity of office space Krier refused and the brief was handed to Quinlan Terry. The latter's scheme was favoured by the planning committee and the community and given the backing of SAVE, the Royal Fine Art Commission and inevitably the Prince of Wales in July 1987. But while Rosehaugh Stanhope succeeded in winning the 'design competition' the Corporation accepted the SDG bid. The wider public remained largely confused about who was going to build what.

The Spitalfields saga provides an extreme example of how the 'battle of the styles' has invaded the commercial realm, of how architectural patronage has become a key part of development tactics, and of the tensions between the highly commercial US practices and the British practices which still have little or no experience of design and development at this scale. It illustrates the inexorable trend towards overdevelopment, especially in the wake of the Bishopsgate phases of Broadgate. But the decline in quality of the design from the 1987 permissions is related to the decline in office demand, the rise in interest rates and the onset of the recession.

CULTURAL INVESTMENT AND URBAN DESIGN –
ROYAL OPERA HOUSE AND SOUTH BANK

In Europe at large few people would look to commercial development to provide the design framework for the new city or to take the lead in providing the architectural foci or new spaces, enhancing the pedestrian network or even providing the transport infrastructure. It is a measure of how much the rôle of the public sector has been transformed by Thatcherism in the last decade that major public investment in urban spaces or cultural facilities has been moved completely off the agenda in Britain. Different traditions of government, different constitutions, different legal frameworks and planning controls notwithstanding, there could scarcely be a greater contrast between the deliberate neglect and emasculation of London practised by the Conservatives and the competition between successive French Presidents to use Paris as a focus for cultural development. Such investments have obvious spin-offs in terms of international economic competitiveness, quality of environment and tourism.

Paris has benefited from an extensive cultural investment in new museums, concert halls, sports facilities, new public buildings and new public open spaces. These have acted to refocus the City's image, revitalize its cultural life and improve its international appeal to commerce. The plans for the eastern city are a model of urban design-led regeneration and while questions must be asked about the intended scale of gentrification, it is no more dramatic than what has transpired in Covent Garden, and markedly less so than will take place in Spitalfields. The fact that Paris now has four modern architectural/urban design museums, including the municipal Pavillon de l'Arsenal ('a device for thinking about the future which is so powerful and dazzling that it makes London seem blind', wrote Frank Duffy) emphasizes the linkages in Paris between public patronage and modern architecture, urban design and cultural investment, and public education that seem non-existent in London.

In London one has by comparison the fiasco of the extension of the National Gallery (first compromised by the requirement that each scheme be 'commercially viable', producing speculative office elements averaging 7,000 square metres), and the long-running saga of the proposed extension to the Royal Opera House designed in 1983 but still not resolved. Although an architectural competition was launched in 1983, and won by an imaginative scheme by Jeremy Dixon, the requirement that the scheme had to be self-financing dictated a large commercial element (at least two times site coverage with a two-thirds office, one-third retail component, to say nothing of a 300-space car park). This severely compromised the listed buildings on the site and although the scheme gained permission from Westminster City Council the resultant overdevelopment aroused the ire of the Covent Garden Community Association, who had the decision successfully reversed by the High Court on the grounds that conservation policies were being ignored. Redesigns ensued in 1989 but generally Dixon's architecture was widely applauded as restoring coherence to the derelict corner of the market, while achieving a creative tension between modernism and classicism. However, an alternative scheme has been worked up by the Covent Garden Community Association and the indecision now stretches over two decades.

Another key cultural facility starved of public investment is the South Bank Arts complex with its nuclei of National Theatre, Hayward Gallery, Royal Festival and Queen Elizabeth Concert Halls, all built in the 1950s and 1960s. These became the management responsibility of the South Bank Board with the demise of the GLC. Reputed to be the largest centre for the arts and entertainment in the world the South Bank is generally regarded as the antithesis of what an urban arts complex should be, by virtue of its monumental, separated, and generally inward-looking elements that fail to relate to the public life of the city and to fully animate the riverside.

The South Bank Board commissioned the architect Terry Farrell to produce some cosmetic improvements in 1985 and he suggested the possibility of much more radical improvements. This resulted in a limited competition amongst developers and the selection of Stanhope Properties (alias Stuart Lipton) with Terry Farrell to promote widespread improvements. The 1989 scheme assumed no public subsidy and involved the developer taking a 150-year lease on the site and investing £200m. in the project in return for 30,000 square metres of office and retail space to the east of Hungerford Bridge. The key to Farrell's plan was a return to ground-level circulation to concentrate and clarify movement, filling the present gaps between the concrete hulks with buildings and public facilities, to recreate the gaiety of the 1951 Festival of Britain and establish a vital public realm. But many critics railed at the 'planning by press release' Gallup poll public participation that constituted Lipton's attempt to lubricate the development control process. In what other capital city would the local government be forced to 'await a planning application' while an unaccountable Board toyed with major national cultural assets? Presumably the recession has put paid to these plans temporarily, but it is the proposed process of consultation and commercial cross-subsidy that is so revealing.

NEW PUBLIC SPACES IN LONDON?

If the lack of cultural provision has been a hallmark of the last decade so too has been the failure to enhance or create new public spaces. While architectural critics remain wedded to the necessity of new architecture as the creator of urbanism European cities have shown that pedestrianization, traffic calming and public transport provision can create both the space and the environmental quality to transform the quality of a city. Munich and Stuttgart are two very fine German examples with extensive footstreets, while Bologna, Vienna, Copenhagen and now Bordeaux provide similar learning models (TEST 1988). These ideas have been absorbed by historic cities like York but they have made little or no impact in London except in Covent Garden. Lobbyists have shown the potential for pedestrianization and environmental enhancement in the City and their proposals would do more to enhance the quality of the environment for the user, and at less cost, than any other investment, policy change or innovation (Eley *et al.* 1986). The completion of a north-side Thames walkway through the City ought to be a major priority (Hillman 1989).

In terms of investment in public space one has to struggle hard to think of new or dramatically remodelled urban spaces outside of Broadgate. Barcelona, by contrast, provides an excellent example of a pro-active approach to urban space by the municipality (in an economic recession) remodelling 160 or so of the city's spaces with dramatic landscaping and furnishing and treating the exercise as a massive commission of public art. It is a piecemeal approach and fails to take the opportunity to impose a new urban structure and clearer image for the city, but it is correspondingly easier to implement and its impact still dramatic. It has had a major positive impact upon the city's international image and was critical to its successful bid for the 1992 Olympics.

In London, since the First World War, there has been a failure to develop new parks, the existing ones remaining primarily a Victorian heritage. Tower Bridge Park and Coin Street in Southwark are two exceptions but both are unimaginatively designed. While the Royal Parks remain adequately funded, at least while they were a DoE responsibility, borough parks have no such resources and even their maintenance is problematic. New parks are planned at the centre of the King's Cross development and in the Greenwich peninsula. In King's Cross Norman Foster's design of the 34-acre park allows the extension of prime office development land much deeper into the scheme by making the sites more attractive and thus more commercially valuable. However, the park could provide an interesting and lively combination of traditional Royal Park with regenerated indus trial architecture and commercial/cultural canalside uses, while tree-lined boulevards would help to provide a structure and environmental quality for this 125-acre site. Apart from various schemes in Docklands and the Greenwich peninsula the possibility of new parks seem remote. Again the contrast with Paris is instructive with its particular programme of open space enhancement in the north-east of the city.

From large-scale traffic calming and pedestrianization through to the remodelling of small open spaces and on to the creation of new pedestrian links and green routes, and the careful integration of urban parks with green corridors of canals, river valleys and woodland linking out into the green belt, there are enormous opportunities to enhance the environment of urban and suburban London. Such investments would promote walking, cycling and other beneficial forms of exercise and recreation and ecologically sustainable transport as well as increasing the accessibility of amenities and enhancing the imageability (the capacity to evoke a strong sense of place in the observer) of the city. There are models in the Groundwork initiatives in northern England or Leicester's ecology strategy, just as there is inspiration to be derived from the more centralized urban design initiatives for Paris and Barcelona, the traffic calming in German cities or the green-space planning in the Netherlands.

LONDON DOCKLANDS – A NEGATIVE AND POSITIVE EPITOME

It is of course in London Docklands that the collapse of strategic planning, the abdication of design control and failure to adhere to even the most elementary principles of urban design have been most dramatically reflected in the urban landscape. The Isle of Dogs Enterprise Zone has been variously described as 'the capital's greatest architectural planning and social disaster', 'the worst collection of late twentieth-century buildings anywhere', 'an architectural zoo' or most graphically a 'Dog's Breakfast'. With even some of the watersides appropriated for private use, the Light Railway snaking and clanging overhead on inelegant concrete viaducts, an almost total absence of usable public space, and the completely random juxtaposition of land uses and building styles, the Isle of Dogs has to be seen to be believed. Now that Canary Wharf is being developed in its midst, with even its lowest buildings exceeding the height of the listed warehouses opposite by four or five times, the effect is positively surreal.

Ironically it was the Canary Wharf project itself which led to a change in the LDDC's *laissez-faire* policy; for G. Ware Travelstead and their architects expected urban design guidelines to enable them to master-plan the site and prescribe continuities of massing, open space and architectural treatment between individual buildings. Such guidelines were accepted practice in North America to protect the investment of each developer/investor, and to ensure an overall quality to the external environment, besides providing the necessary functional linkages from scheme to scheme (Buchanan 1989a). It was left to Skidmore Owings & Merrill to develop the guidelines for Canary Wharf, but the LDDC subsequently commissioned guidelines for Heron Quays to the south in a similar guise and adopting a similar Beaux Arts symmetry. Subsequently these have been superseded by new urban design guidelines commissioned by the actual developers Olympia & York. Experience gained at Canary Wharf has encouraged the LDDC to embark upon design briefs for East Beckton, the Royal Docks and many other major sites.

In Surrey Docks a landscape master plan was prepared as far back as 1982 and has helped to provide a strategic framework for urban design and open-space planning. In the Royal Docks Richard Rogers Partnership were retained to design an infrastructure system and to broadly differentiate different intensities of development and provide a landscape framework. The site was then divided into large parcels for individual developers to put forward their own master plans and financial offers. These verbal and graphic guidelines were refined into legal documents by consultants in consultation with the LDDC, specifying the nature and finishes of the highway network, view corridors, building lines, some uses, landscaping and public spaces. These guidelines and a landscape master plan are no panacea, and offer only the most basic frameworks for development commensurate with the LDDC's general *laissez-faire* ideology, but they are ensuring the creation of a public realm, reserving the waterfronts for public use, and suggesting an appropriate scale and relationship of developments.

So, paradoxically, London Docklands now offers in elementary form precisely the kind of urban design framework that ought to guide the redevelopment and enhancement of the rest of London, providing a relationship between infrastructure and density, ensuring the continuity of the public realm and public access to key amenities like the quaysides, and ensuring a linked landscape framework that can provide an ecological and recreational amenity and the coexistence of different architectural treatments and land uses. This should prevent a repeat of the visual and functional discontinuities of the Isle of Dogs.

Finally it should be recorded that Docklands contains more fine 1980s buildings than any other comparable area of a British city (e.g. Rogers' and Outram's pumping stations, Grimshaw's *Financial Times* building, Dixon's housing and Troughton McAslan's apartments spring instantly to mind), and a significant amount of good conservation (Williams 1990). Shad Thames and St Saviour's Wharf combine the two with the restoration of the Anchor Brewhouse and Butler's Wharf cheek by jowl with Wickham Associates' Horselydown Square, Conran Roche's and Michael Hopkins' modernist buildings, and CZWG's sculptural and colourful forms. Here there are signs of a new urbanism of quality. But as one celebrates this new townscape so one is reminded of the inability of London Docklands Development Corporation to grapple with the social dimension of urban regeneration and its inability to ensure an appropriate social mix in its developments. The narrow mandate (particularly the inability to subsidize land disposals), the pernicious boundaries, hostile and impoverished local authorities, and now a general shortage of funds have all conspired to enforce gentrification on Docklands and ensured its failure to integrate the local community within the development success story. Will Docklands ever be more than a chain of highly polarized ghettoes epitomizing the gulf between rich and poor, home-owner and tenant? Will it continue to be a place apart from the rest of London – distant, fragmented, unintegrated and introverted, just as it was in the past?

A VISION OF LONDON – A RESPONSE TO THE CRISIS

The crisis of architecture and urban design in London has no simple remedies. The reasons for this lie in the fact that it is not possible to separate urban or civic design from questions of political authority and process, strategic and local planning vehicles, infrastructure provision and externality costs, social content and betterment. Central government of both political hues has drifted or abdicated in each of these areas, initially because of a lack of political courage, and more recently because of political ideology and the belief that the market will produce the best solution. Most of the problems for urban design revolve around the question of profit, whether they be questions of the density, social content, amenities, infrastructure or other planning gains or responsibilities. Profit is directly related to the risk which is borne by the developer: the greater the risk the greater the potential profit (and loss). The issues of profit are not ostensibly planning or design issues; but local authority determination to share development profits in one form or another has a major impact on design. This determination has been largely generated by central government refusal to fund even the most basic infrastructure, and by tight controls on all local expenditure which preclude most forms of environmental enhancement (and now even adequate staffing levels). At present developers are expected to provide contributions to 'infrastructure' of many different kinds – strategic networks, infrastructure specific to the project (usually linkages to strategic infrastructure), facilities to reduce the direct externalities of the development, like congestion, and amenities appropriate to good planning. These are all negotiated against the background of the land cost, and the expectations of landowners are geared to current rents and plot ratios. In any sane planning system the strategic infrastructure must be provided by the state, the specific infrastructure preferably related to standardized costs, the externality facilities related to an analysis of environmental impact, and the amenities resolved by design briefing and control. Otherwise we are reduced to the situation where Olympia & York are planning the new Underground system, Stanhope is deciding the Channel Tunnel terminus and link route and the cultural content of the South Bank, leaving LET and others to relocate the city's markets. The results of leaving the future of London to market processes are there for all to see.

The solutions to the Gordian knot of land values, infrastructure costs, land-use mix, densities and design quality lie in a development land tax or a properly systematized form of planning gain which can be directed to key infrastructure. Or they lie in public–private partnerships such as those embodied in the *Zone d'Aménagement Concerté* in France, where the local authority acquires the land at existing use value and works in partnership with private, and some public, developers to an agreed master plan. The development partnership provides the key infrastructure and design framework, writing down the land values of socially necessary uses like hospitals, social housing or open space but providing plenty of opportunities for commercial developers to take sites and to respond to the market. New forms of public–private partnership are essential in London not merely to resolve the conflicts over the environmental costs and social content of development, but to provide developers with a high degree of certainty and a more realistic timetable for development, and to ensure a proper level of investment in the public realm.

On the matter of design control itself a much more positive attitude is required by central government, beginning with the rewriting of Circular 22/80, which tried to argue that design control was only allowable in conservation or other 'environmentally sensitive' areas. Beyond that there is a need for a strategic plan which sets a framework for urban design, specifying the capital web (the open-space system, mass-transit lines, the highway hierarchy), key nodes for intensification and redevelopment, and areas for conservation and protection. Then there is a need to develop more detailed guidance in local plans and development briefs. Such guidance is often well developed for Conservation Areas but needs to be evolved for major redevelopment zones. Many architects and urban designers have pointed the way – MacCormac, Tibbalds, Duffy, Farrell, Buchanan among others – and experience gained in the mega-developments already outlined will be invaluable in improving the quality of design briefing. Density remains one of the key issues and the blanket introduction into the City of a 5:1 plot ratio can now be seen to be a retrograde step that is promoting overdevelopment on many sensitive sites. A more selective application is necessary and more attention needs to be directed to the rules for air rights and basements in order to prevent repeats of the Bishopsgate phase of Broadgate, or even crasser architectural forms like Beaufort House or Minster Court in the City.

Most important of all is a focus on the public realm, the creation of permeable developments that retain some of the grain of the historic city and which express the main pedestrian desire lines. A hierarchy of routes needs to be established – arterial, vehicular, public and semi-public, covered and uncovered. Pedestrian routes need to be lined, especially at the ground level, not with vast offices but with 'local', smaller-scale functions that are used by a wide community – small offices and services, shops, catering, leisure, housing of all tenures, etc. Architecturally the ground floors need to be invested with especial richness to enhance the pedestrian experience. Spaces should have an enhanced micro-climate, a human scale, appropriate furnishing and landscaping and be related to the functions of the area. A thoroughgoing contextualism which integrates morphology, social use and architecture can provide the necessary framework for development without stifling architectural innovation.

Issues of architectural style come last in terms of priority, and it has been an unfortunate aspect of the debate about the quality of the new urban environment over the last decade that they have come first, because it has deflected concern from the content, equity and user experience of development. But issues of style are still important, especially when much broader and taller buildings are being inserted into the city. How to treat such huge expanses of elevation, especially in highly visible and historic locations is an important issue. It is not resolved by advocacy of particular styles but by insistence on certain principles of public–private, visual and functional interaction through the façade, appropriate scaling, choice of materials and appropriate detail and decoration (Buchanan 1988b). Stylistic pluralism, innovative contextualism, and design quality must be the watchwords. A more open, informed and less polarized debate, with wider public participation and consultation, would mean that the lay persons could speak for themselves instead of having the Prince of Wales do it for them. Safe streets, affordable housing, lower density, small spaces, active frontages, reliable public transport and abundant greenery would be the foci of the design agenda, not skyline protection or the advocacy of particular architectural styles. Mean streets are more important negative factors in London than carbuncles of all kinds.

The search for *A Vision for London* such as that launched by the Association of London Authorities in March 1990 (*Architects' Journal* 1990) will demand workable solutions to issues of strategic planning, infrastructure provision, planning gain and design control. It will demand a democratic decision-making process in which the design professions – architects, planners, landscape architects, construction managers – will all have to play a part as sources of workable ideas and implementation expertise. It will demand outstanding political leadership and political courage at the national, London-wide and local level. It will demand a major reinvestment in the public realm, especially in public transport and public space. And last but not least it will demand the creativity and determined patronage of developers keen to leave positive contributions to humanity. Curiously there is some evidence of the latter in contemporary London but precious little of any of the former.

REFERENCES

Books

Booker, C. and Lycett Green, C. (1973) *Goodbye London: An Illustrated Guide to Threatened Buildings*, London, Fontana.
Duffy, F. and Henney, A. (1989) *The Changing City*, London, Bulstrode Press.
Glancey, J. (1990) *New Directions in British Architecture*, London, Thames & Hudson.
Hillman, J./Royal Fine Art Commission (1989) *A New Look for London*, London, HMSO.
Jencks, C. (1988) *The Prince, the Architects and New Wave Monarchy*, London, Academy Editions.
Williams, S. (1990) *Docklands*, London, Architecture, Design and Technology Press.
Williamson, R. J. G. (1989) *A Discipline for Change in the City of London and Environs*, Oxford, Oxford Polytechnic Joint Centre for Urban Design (M.A. thesis).

Journal Articles

Architects' Journal Eds *et al.* (1990) 'A vision for London', *Architects' Journal* 191, 14 March, pp. 29–86.
Buchanan, P. (1988a) 'What city? A plea for place in the public realm', *Architectural Review* 184 (1101):31–41.
—— (1988b) 'Facing up to façades', *Architects' Journal* 188, 21 and 28 December, pp. 21–56.
—— (1989a) 'Quays to design', *Architectural Review* 185 (1106):39–44.
—— (1989b) 'Paternoster pressure', *Architectural Review* 185 (1107):76–80.
Cruickshank, D. (1986) 'City within a city (Spitalfields)', *Architects' Journal* 184, 5 November, pp. 12–15.
—— (1989a) 'Classicism and commerce (Paternoster)', *Architects' Journal* 190, 15 November, pp. 28–31.
—— (1989b) 'Market options (Spitalfields)', *Architects' Journal* 190, 25 October, pp. 26–31.
Darley, G. (1988) 'No news is bad news (Fleet Street)', *Architects' Journal* 187, 10 February, pp. 24–9.
Davey, P. (1988) 'Architecture v. big business', *Architectural Review* 184 (1098): 15–19.
Davies, C. and Duffy, F. (1988) 'Unbuilt London, Paternoster: Bracken: Stag', *Architectural Review* 183 (1091):15–38.

Eley, P., Ward, B. and Smith, G. (1986) 'Building for the Square Mile', *Architects' Journal* 182, 9 April, pp. 28–49.

Farrell, T. (1987) 'Terry Farrell Partnership: current London projects', *Architectural Design* 57(1/2):44–53.

Foster, N. (1990) 'Norman Foster: King's Cross: a master plan', *Architectural Design* 60(3/4):36–43.

Jencks, C. (1988) *The Prince, The Architects and New Wave Monarchy*, London, Academic Editions.

MacCormac, R. (1987) 'Fitting in offices', *Architectural Review* 181 (1082):62–7.

Marmot, A. F. and Worthington, J. (1987) 'Great fire to Big Bang: private and public designs on the City of London', *Built Environment* 12(4):216–33.

Rabeneck, A. (1990a) 'The American invasion', *Architects' Journal* 191, 28 March, pp. 36–57.

Rabeneck, A. (1990b) 'Broadgate and the Beaux Arts', *Architects' Journal* 192, 24 October, pp. 34–51.

Spring, M. (1988) 'Stretching City limits (Broadgate)', *Building* 14 October, pp. 41–8.

TEST (1988) 'Quality Streets', London, TEST.

Tibbalds, F. (1990) 'Spitalfields inspiration', *Building Design* 12 January, pp. 14–17.

Weekend Guardian eds (1989) 'Six buildings in search of an architect', *Weekend Guardian*, 25 February, pp. 2–4.

Welsh, J. (1989) 'Waterworks (Docklands)', *Building Design* 27 October, pp. 40–5.

Worthington, J., Franks, M. and Spence, A. (1986) 'City speculations', *Architects' Journal* 183, 9 July, pp. 27–40.

7

LONDON'S STREETS OF FEAR

Gill Valentine

On 29 December, Alison Day was sexually assaulted and murdered after leaving Stratford railway station, in East London, to meet her boyfriend; her body was dumped in a canal near Hackney Wick station.

This is just one example of the personal stories which make up the Home Office crime figures each year. And these statistics say that in the late 1980s there has been a marked increase in the number of violent crimes reported by women. In 1989 3,305 rapes were reported, compared with 1,842 in 1985, and a total of 15,370 cases of indecent assault were recorded during 1989, compared with 11,410 in 1985.

In London the total of 963 rapes reported to the Metropolitan Police between October 1989 and September 1990 was a 10 per cent rise on the previous year. This reflects an estimate in the *Guardian* (Campbell 1990) that about a fifth of all crime in Britain takes place in London. There was also a large increase of 14 per cent in the number of cases of indecent assault reported against females, taking the total to 2,800. And in 1989 alone three rapes were reported on the London Underground. But if the London Rape Crisis Centre is to be believed, these figures are just the tip of the iceberg. It estimates that only 25 per cent of women who ring the line actually report their rapes to the police (London Rape Crisis Centre 1984), and it claims that the number of women contacting it has increased by between 20 and 40 per cent since 1985. Against this background it is hardly surprising that for many women London's streets are paved not with gold but with fear.

And despite the fact that statistically men are more likely to be victims of crime than women it is well established in the sociology and criminology literatures of western Europe that women are the gender more fearful of crime, and that this is related to women's sense of physical vulnerability to men, particularly to rape and sexual murder, and to an awareness of the seriousness and horror of such an experience (Baumer 1978; van Dijk 1978; Toseland 1982; Valentine 1989a).

Sociologists researching public perceptions of crime have also consistently reported a strong correlation between fear of crime and city size (Toseland 1982). This relationship is understood to be a consequence of the generally higher crime levels that characterize large urbanized areas such as London, which mean more people are likely to be exposed directly or indirectly to crime (Boggs 1971). It is due also to the fact that the greater range and complexity of the social worlds which characterize larger places cause uncertainty and anxiety about the nature of other people who may be present, and so increase suspicion and fear of others (Kennedy and Silverman 1985).

For the majority of women London therefore is perceived as a city of danger. But little has been written about how this fear increasingly affects women's freedom of movement within and experience of life in the city.

Therefore in the first part of this chapter I will draw on my Ph.D. research to explain the relationship between women's fear of male violence and their perception and use of public space.[1] Although the theory I will set out is based on qualitative research conducted in Reading I believe that the findings are applicable to women living in other western European towns and cities and this belief is borne out by the results of London surveys which I will use to support my theory. In the latter part of the chapter I will discuss initiatives past and present to make London a city free of fear.

PLACING FEAR ON THE STREETS

Concern surrounding the sexual assault or murder of women often centres more on the victim's use of space than on the nature of the men who commit these crimes.

Alison Day was in an isolated public place away from the protection of others when she was attacked. Both the police and the media tend to imply that such women are to a certain degree responsible for their own fates by putting themselves in these situations, and they warn other women to avoid placing themselves in similar situations of vulnerability.

This public blame of victims for being in 'dangerous' or inappropriate places when they are attacked encourages all women to transfer their threat appraisal from men to certain public spaces where they may encounter assailants (Valentine 1989a).

The other side of this fear of being in public space is that women falsely perceive themselves to be safe in the private space of the home.

The A–Z of women's fear

Women develop individual mental maps of places where they fear assault as a product of their past experience of space and secondary information. In particular girls are socialized into a restricted use of public space through observing both their parents' differential fears for them and the control of the spatial range of their activities in relation to boys (Hart 1979). Consequently most girls have mental images of places where strange men may approach them instilled at an early age.

However, despite their fears and their possible avoidance of 'dangerous places' most women do have some form of frightening experience such as being flashed at or followed (Kelly 1987; Valentine 1989a). A 1984 survey carried out by Ruth Hall in London found that of the 1,236 respondents to her questionnaire, at some point in their lives 83 per cent had been verbally assaulted, 76 per cent physically grabbed, and 72 per cent followed or flashed at (Hall 1984). And a survey carried out on behalf of the GLC Women's Committee in 1984/5 found that of the 905 women interviewed 15 per cent of women living in Inner London and 9 per cent of women in Outer London had been harassed in the last year (GLC 1985). Such incidents then become associated with the environmental context in which they took place, so reinforcing or developing women's geography of fear.

Additionally these mental maps of feared environments are elaborated by images gained from hearing the frightening experiences and advice of others (Valentine 1989a) and from the media. The GLC survey found that of the women interviewed 39 per cent of those living in Inner London and 23 per cent of those from Outer London knew of another woman who had been harassed in the last year. And Morley's (1986) study of television and gender revealed that women watch more local news and police programmes than men because 'if there has been a crime (for instance a rape) in their local area they need to know about it, both for their own sake and their children's sake' (Morley 1986:169).

Women assume that the location of male violence is unevenly distributed through space and time. In particular women learn to perceive danger from strange men in public space despite the fact that statistics on rape and attack emphasize clearly that they are more at risk at home and from men they know. This is because when in public the behaviour of any stranger encountered is potentially unpredictable and uncontrollable (in this context my research suggests that women perceive only men and not other women as strangers).

The particular types of places in which women feel most at risk are therefore where they perceive the behaviour of men to be unregulated. First, large open spaces which are frequently deserted: parks, waste-ground, and rivers. Frequently local place mythologies develop around such places. Burgess's open space project in Greenwich found an association between wooded parks and 'dirty old men' (1987).

Secondly, there are the closed spaces with limited exits where men may be concealed and able to attack women out of the visual range of others: subways, alleyways, multistorey car parks, and empty railway carriages. Such opportunities for concealed attack are often exacerbated by bad lighting and ill considered building design and landscaping (Heing and Maxfield 1978).

For example 73 per cent of women questioned in an East London survey conducted by Kate Painter of Middlesex Polytechnic said they were afraid to use Watney Street, which is an essential route between the Dockland Light Railway and housing estates. And respondents said they ran from the railway exit or deliberately crossed the street to avoid the most badly lit area. But after an experiment to improve the street lighting 94 per cent of people questioned said their fear of crime had been reduced (Souster 1989).

Unequal spatial opportunities

This association of male violence with certain environmental contexts has a profound effect on many women's use of space. Every day most women in western societies negotiate public space alone. But many of their apparently taken-for-granted choices of routes and destinations are in fact the product of the 'coping strategies' they adopt to stay safe (Stanko 1987, Valentine 1989a).

Women adopt three different strategies to minimize their time alone in public space. First, many women simply reappraise their need to go out, either by 'choosing' to stay in or by denying to themselves that they want to go out. This is the strategy most commonly adopted by older women whose options are limited by reduced strength, health and agility and diminished financial resources.

Second, women avoid being in public space by travelling between destinations in the private space of their own and others' cars. This coping strategy is most frequently adopted by middle-class married and single women in full-time employment who have access to private transport. For example the 1984/5 GLC survey found that 40 per cent of women interviewed in full-time paid work used a car at least once a week, and 39 per cent of those working part time, compared with only 26 per cent of those not in paid work.

Finally, women avoid being in public space alone by going out in the evening with a group of friends or parents and partners. This coping strategy is commonly adopted by young women with extensive local friendship networks but few other 'safety resources' such as cars or cash for taxis. Their plight has been exacerbated by the end of the Fares Fair policy which had kept down fares on London transport, and by transport cuts which have reduced the number, and destinations, of train services.

When women are forced by circumstance to be in places where they feel at risk they become alert to their physical and social surroundings, listening for rustles in the bushes or approach of footsteps. As a result most women, especially at night, have a heightened awareness of the micro-design features of the environment and adjust their pace and path accordingly: running past or crossing the road to avoid alleyways, indented doorways, and shadowy areas.

Besides these adjustments women also change their physical appearance and possessions in order to deter assailants, to help them escape from threatening situations and to resist assault. Consequently as a product of their fear many women not only perceive but also experience London differently to men.

Social control, space and time

Not all public places are perceived as equally threatening all the time, because in many places or at some times the behaviour of those occupying the space is externally regulated either formally or informally, so reducing the perceived oppor-tunity for attack. Formal control of public space is exercised not only directly by the police or private security guards, but more indirectly by store managers, bus con-ductors, and other authorized personnel in the process of providing a public service.

However, in the last decade transport cuts, and the privatization of bus services, have resulted in the introduction of one-person-operated buses and trains and unstaffed stations – factors which have undermined women's safety in these previously regulated environments. The number of attacks on people on buses steadily increased during the 1980s; in 1984 there were 1,176 attacks. And the GLC survey (1985) shows that women prefer crew-operated buses as opposed to those operated by one person (OPO), with 55 per cent of those questioned saying they would rather wait for a crew-operated bus, if they had the choice, than get on an OPO bus.

Informal social control in public areas relies upon the potential intervention of others present to act as a deterrent to those contemplating crime. This control is more successful in stable neighbourhoods where people have strong social and family ties and are therefore more easily able to recognize strangers and are more likely to feel confident to intervene to help others, or to know where to seek help if they feel threatened (Riger and Lavrakas 1981).

However, in housing estates where there is a high turnover of population and few community resources the inhabitants are often strangers to each other and the place, and therefore there is often little if any informal social control. In parti-cular a common problem on London tower block estates is the low morale of the residents, the anonymity of the structures, and poor design, which all contribute to reducing the ability of occupants to control common space (Southwark Borough Council 1984).

A woman's perception of her safety in her local neighbourhood is therefore strongly related to how well she knows and feels at ease with both her physical and social surroundings.

When a woman is in an area beyond her local environment she makes judge-ments about her safety in public space on the basis of preconceived images she holds about the area and its occupants, as well as cues she receives about social behaviour from the actual physical surroundings. Fears of potential hostility are particularly centred upon vandalized places, areas containing graffiti, and residential areas identified on the basis of ethnicity and class such as Brixton (Valentine 1989b). The increase in the number of homeless and mentally ill on the streets as a result of social policies and spending cuts has also added to women's concerns about the nature of groups dominating public space.

Beyond an attribution of control to the major residential group, the group which is actually dominant in a public space fluctuates with time of day according to gender and age. During the daytime public places are numerically dominated by women in part-time paid work, housewives, young children and the elderly. This is because of their limited access to private transport, flexible time budgets and need to fulfil domestic tasks such as shopping. Those men who are present are usually dressed in work clothes and engaged in work-related activity and therefore their behaviour appears both predictable and controllable.

As evening draws in, it is younger people, and particularly men, who are visible. Freed from the confines of work, and usually without the family responsibilities of most women, they have the time, energy and financial resources to go out in the pursuit of leisure activities and therefore to numerically dominate public space. Consequently, whilst women identify specific isolated places as frightening during the day, they express a fear of all public space alone at night. The GLC survey (GLC 1985) reported that 63 per cent of women interviewed avoid going out alone at night. Thirty per cent strongly agreed with the statement 'I don't go out on my own after dark', and only 15 per cent feel safe walking alone at night. This is not only because night reduces visibility and therefore increases the opportunity for attackers to strike unobserved, but because the nature of public space changes, being dominated in the evening by the group women most fear, men.

Spatial expression of patriarchy

Women's fear of male violence does not therefore just take place in space but is tied up with the way public space is used, occupied and controlled by different groups at different times. There is a vicious circle in operation. The majority of women still adopt a traditional gender rôle, and as a consequence are pressurized into a temporally segregated use of space. The subsequent control by men of public space in the evening means that despite the career success and independence gained by some women in the past decade the fear of male violence deters the majority of women from being independent. It robs them of the confidence to live alone, to work in certain occupations and to socialize without a group or a male chaperon.

This inability of women to enjoy independence and freedom to move safely in public space is therefore one of the pressures which encourages them to seek from one man protection from all. This dependence on a single man commonly limits women's career opportunities and general life-world. This in turn results in a restricted use of public space by women, especially at night, allowing men to appropriate it and hence making women feel it is unsafe to go out, reinforcing their comparative confinement in the home. Consequently this cycle of fear becomes one subsystem by which male dominance, patriarchy, is maintained and perpetuated. Women's inhibited use and occupation of space is therefore a spatial expression of patriarchy.

TOWARDS A LONDON FREE OF FEAR?

'I don't like being victimized and I think that's what it is, you are victimized because you're a woman. I mean something's got to be done because women can't even go shopping. At Asda's there was a woman raped a while ago. You should be able to go shopping without wondering whether you're gonna get raped. So somebody's got to do something soon'

(young woman quoted in Valentine 1989a)

This plea has not fallen totally on deaf ears. I now want to go on to use the understandings about the relationship between women's fear and space outlined above to suggest what can be done to improve women's safety in London: first, by briefly considering what can be done to prevent violence against women; second, by examining what can be done to make women more confident and able to defend themselves; and finally, by considering the initiatives taken by London authorities in response to what women think would make them feel safer.

What can be done to stop male violence

The weaknesses in the sociobiological theories which have been used to argue that men are driven to rape either by a 'natural' urge to reproduce, or by the influence of hormones (Barash 1977) are well established in the feminist, sociology and biology literatures (Sayers 1982; Birke 1986).

Instead it has been clearly established by sociologists and psychiatrists working with sex offenders that men commit sexual crimes against women for social reasons of power, dominance, anger and revenge (Perkins 1986; Wyre 1986). Furthermore, male violence has been identified as existing on a broad continuum of male power, including not only 'massive' forms of violence such as rape, but also sexual harassment and pornography (Kelly 1987). Violence has also been linked to broader psychological and economic forms of control. But the main focus of feminist concern and protests by groups such as London Women Against Violence Against Women (WAVAW) has become the relationship between sex, violence and power within the social construction of masculinity and heterosexuality (Coveney *et al.* 1984).

Consequently if women are to be free from the fear of male violence we need to explore and challenge the way in which gender identities are constructed and reconstructed.

But although tackling the reasons why men attack women is beyond the scope of this chapter, women can take positive action to redefine femininity by learning to defend themselves and so reduce their fear of male violence.

Increasing women's personal safety: self-defence

Women are afraid of male violence because they perceive themselves as physically unable to resist male assailants (Valentine 1989a). This illusion of helplessness is socially constructed (*Matrix* 1984). In contrast to boys, girls are not encouraged to develop physical strength, speed or fighting skills. Furthermore because women are not encouraged to learn techniques to defend themselves they assume that because on average they are physically smaller than men, that they will not be able to resist a male assailant. This perceived inability to control personal interactions with strange men pressurizes women into a restricted use of public space. Therefore in order to prevent women's fear of male violence operating through space in this way, women must learn to personally control their interactions with men.

To this end women's self-defence classes have been developed in most London boroughs to teach women simple techniques and strategies that can overcome superior physical strength and to help them become confident about their right to be out alone in public space after dark.

However, the growth of these self-defence classes has been restricted by a lack of funding. This is due, first, to the abolition of the GLC, which gave grants for such courses; second, it is because women's units, which were a source of self-defence training, have been scrapped as part of local authority spending cuts. Ealing's women's unit is just one of the victims of such cuts.

And although the publicity surrounding women's fear of crime has resulted in the spread of private self-defence classes in London, many of these are not only expensive but also do little more than teach women limited martial arts techniques. Unless such courses also build up women's assertiveness and reduce inhibitions about using their strength, women are unlikely to be able to exercise the techniques in a confrontational situation.

Finally, the spread of women's self-defence skills in London is limited through some women being reluctant to attend courses because they do not recognize their value, while others are unable to take part owing to domestic or paid employment commitments.

Initiatives to make women feel safer

A lack of self-defence skills means most women look to external forces to make them feel safe in public space.

As outlined earlier, women's perceptions of fear relate to the way public space is designed, occupied and controlled, and as a consequence the strategies they would like to see adopted to improve their safety are directed at changing the way space is organized, occupied, and controlled.

Although changes in the design and layout of the physical environment cannot prevent attacks happening, women perceive that they can reduce the opportunity for them to occur (Valentine 1990). Therefore environmental improvements can increase women's confidence to go out. The ten major design strategies research has shown women would like to see adopted by future planners and developers are listed in Table 7.1 (Valentine 1989a).

During the mid- and late 1980s a number of London borough councils took such practical design measures to try to improve women's safety on housing estates. Many of these initiatives have their origin in pressure from female tenants for action to improve degenerating estates. These demands were made through local council women's units and resulted in working groups being set up

Table 7.1 Design changes to improve women's safety

Location The positioning of car parks and entrances so that women do not have to walk down passageways to gain access to sites

Visible doors Communal entrance and doorways to lifts, stairwells, etc. should be glazed so that women can see through them before entering.

Lighting Bright white rather than yellow lights should be used to maximize visibility

Painting of walls Walls painted white, for example in multistorey car parks, improve visibility and make the space appear more open

Bridges Where possible foot bridges should be constructed rather than subways

Alleyways/subways If these are absolutely necessary they should be as short and as wide as possible, with no overhanging vegetation, and provide maximum opportunity for surveillance from surrounding buildings. The entrance should be clearly visible in advance from the main pavement

Landscaping Trees and shrubs should be planted away from pathways. Similarly, fences and walls should be minimized so that public areas are not screened from houses. Mounds and clumps of trees in recreational areas should be set well back from the edge because they obscure vision of the playing area from surrounding roads

Ground floor development Where possible shops and leisure facilities should occupy the ground floor of office buildings so that alienating empty streets are not created

Fill-in Attempts should be made to clear derelict areas and to fill in gaps, empty spaces and waste areas between developments

Corners, dog-leg bends Straight sight lines should be created where possible. If blind corners are necessary, mirrors could be used to increase visibility. Similarly, doorways should be flush with the building rather than creating indentations where people can be concealed

---

in many London boroughs to look specifically at women's design and safety needs. Other boroughs' initiatives arose as a direct result of violence against women.

In 1986 women from the Honor Oak estate in Lewisham wrote to the council demanding action after a woman was raped and murdered while walking along an isolated footpath on the estate. In response the council improved lighting and fencing along the route, and then consulted tenants about other improvements on the estate. Similarly, the murder of a female tenant on a Southwark estate led her neighbours to make a radio programme called *Women's Safety on Housing Estates*. A copy of the tape was sent to Southwark Women's Committee and in turn this resulted in the council taking action in the form of design changes.

Most of the London boroughs have taken similar actions to improve women's safety since 1984. Each borough went through different consultation procedures.

Hackney conducted interviews with one hundred female tenants, and canvassed other women's opinions at an open meeting on one estate. Lambeth held a day of workshops about safety during which questionnaires were circulated.

The reports produced individually by the boroughs all recommend similar design changes. All of them have taken initiatives to improve lighting on estates; to introduce entry-phone systems to flats; to introduce resident caretakers to carry out foot patrols; to improve locks and doors; and to take into account safety design needs in future investment programmes. Additionally Lambeth has also taken action to remove bridges on two estates, and to block alleyways in fifteen estates.

However, there is little information on the success of these improvements because a number of the schemes are still being put into action; and owing to staffing and structural changes most of the authorities do not have the resources to monitor the progress of these initiatives.

Women surveyed also mentioned that they would feel safer if public places were properly maintained and kept free of graffiti and vandalism. Although the introduction of the poll tax and the privatization of local government services, such as refuse collection, have put pressure on local authorities to cut their spending on such operations, London Transport has made a concerted effort to clean up and redecorate its stations.

As part of the revamp of its stations London Transport also installed video surveillance systems at a further 37 stations in 1986 at a cost of £750,000. Women feel more confident in public spaces which are under surveillance in this way, or which are patrolled by people in authority, because they believe such measures act as a deterent to potential assailants and mean someone will inter vene to help them if they are attacked (Valentine 1989a).

However, despite the introduction of video surveillance, as I outlined above, in the last decade public spending cuts have reduced the number of people employed in public service rôles who fulfil this function, such as ticket collectors on buses and trains. And despite the recent recognition by American and British police forces of the need to return to community policing as part of a conscious strategy to reduce fear of crime (Moore and Brown 1981), a lack of personnel means the police are physically unable to patrol large areas of public space.

As a result, over the past few years there has been a notable growth in private security firms employed to patrol privately owned public places such as shopping malls and car parks; and to guard private housing estates. But this has caused alarm amongst the police because these security guards are not publicly account-able for their actions and have no legal status beyond that of an ordinary citizen. The arrival in London of the Guardian Angels, a vigilante-style force which patrols the New York subways, to train British volunteers to work on the Under-ground was welcomed by passengers but heavily criticized by police.

But the downside of these initiatives to make women feel safer is that agents of control such as the police and private security have been heavily criticized for disproportionately harassing and moving on groups of young men and Afro-

Caribbean and Asian youths, so reinforcing racism and undermining individual privacy (Kelling 1987).

Safe transport is therefore a key to improving women's safety without undermining the freedoms of others. As I have already outlined, the GLC Women's Committee survey revealed that fear of attack whilst on public transport, especially at night, was a major concern of most of the London women (GLC 1985).

A number of suggestions for improvements were made by the women interviewed. These included: women-only carriages, the introduction of closed circuit television on the Underground, reduced fares to increase the number of women using public transport, higher levels of staffing on trains, buses and platforms, improved lighting at bus stops and Underground tunnels, better information about bus and train times and delays, and community minibus schemes.

But during the preparation of this report the Greater London Council was abolished and many of these proposals were lost. Two survivors of the abolition were Safe Women's Transport and the Stockwell Lift Service, local women-only minibus and taxi schemes established in 1982 by the GLC. Funding for the free Stockwell Lift Service was taken on by Lambeth Borough Council, and Lewisham Borough Council assumed responsibility for Safe Women's Transport, a cheap night-time minibus service. A similar scheme for women from ethnic minorities was also introduced in Camden.

The Centre for Independent Transport Research in London is also aware of women's safety needs on public transport. It campaigned throughout the late 1980s (the Campaign for Improved London Transport) against the introduction of one-person-operated trains and buses and against the introduction of unstaffed tube stations.

The Home Office has also launched a number of practical schemes, including advertising campaigns and leaflets such as *Positive Steps for Women* (Home Office 1987) offering advice about women's safety. In 1982 the Neighbourhood Watch scheme was introduced with the aim of reducing crime and fear of crime by encouraging informal social control. It was widely taken up both in London and nationally. And in 1988 the government also launched the Safer Cities programme with the aim of reducing fear of crime. It encourages councils to establish 'Fear of Crime Reduction Committees', composed of the police, probation service, and local authority, commercial and community representatives, to develop local initiatives to tackle fear of crime. However, this scheme is hampered by lack of resources with central government holding the purse strings.

Like the London borough councils, the government has also been concerned about the relationship between environmental design and crime and fear of crime. Recently, the Home Office Crime Prevention Training Centre has produced the *Manual of Guidance for Police Architectural Liaison Officers*. This aims to provide guidelines for chief planning officers designing new environments which will reduce opportunities for crime, and so reduce fear of crime.

CONCLUSION

The 1980s was a decade of rising fear of crime for London women. During this period there was a significant rise in reported crimes of violence against females. And their perception of safety in public space was further undermined by public spending and transport cuts which reduced the regulation and maintenance of public places and transport services, making women feel they are more vulnerable to potential assailants in public. Furthermore social policies have increased the number of homeless and mentally ill living on the streets of London; this has increased women's distrust of the type of people they may encounter in public places at night.

In the mid-1980s the Greater London Council and London borough councils funded research and practical projects to understand and tackle women's fear of male violence in London. But the abolition of the GLC and the reorganization of local government taxation and funding have cut off the funds for, or brought to a premature end, many of these programmes.

At the beginning of the 1990s for women, London still seems a city of danger.

NOTE

1   This chapter is based on the results of 80 in-depth interviews (with accompanying spatial diaries) and six small-group discussions with Reading women of varied age, lifestyle and income. In addition periodic recorded observations of specific public space were made. The full findings are published in the Ph.D. thesis (Valentine 1989a) referenced below.

REFERENCES

Barash, D. P. (1977) *Sociobiology and Behaviour*, New York, Elsevier.
Baumer, T. (1978) 'Research on fear of crime in the US', *Victimology* 3:254–67.
Birke, L. (1986) *Women, Feminism and Biology*, Brighton, Wheatsheaf Books Ltd.
Boggs, S. L. (1971) 'Formal and informal crime control: an exploratory study of urban, suburban and rural orientations', *Sociological Quarterly* 12:319–28.
Burgess, J. (1987) Unpublished paper presented at Reading University Seminar, 1 December.
Campbell, D. (1990) 'Victims readier to report rape attacks, police say', *Guardian* 16 November.
Coveney, L., Kay, L. and Mahony, P. (1984) 'Theory into practice: sexual liberation or social control', in L. Coveney, M. Jackson, S. Jeffrey, L. Kay and P. Mahony, *The Sexuality Papers*, London, Hutchinson, 85–103.
GLC (Greater London Council) (1985) *Women on the Move* (Part 4), London, GLC.
Hall, R. E. (1984) *Ask Any Woman: A London Inquiry into Rape and Sexual Assault*, Bristol, Falling Wall Press.
Hart, R. (1979) *Children's Experience of Place*, New York, Irvington.
Heing, J. and Maxfield, M. (1978) 'Reducing fear of crime: strategies for intervention', *Victimology* 3:279–313.
Home Office (1987) *Positive Steps for Women*, London, HMSO.
Kelling, G. (1987) 'Acquiring a taste for order', *Crime and Delinquency* 33(1):90–102.

Kelly, L. (1987) 'The continuum of sexual violence', in J. Hanmer and M. Maynard (eds) *Women, Violence and Social Control*, Basingstoke, Macmillan, 46–60.

Kennedy, L. W. and Silverman, R. A. (1985) 'Perceptions of social diversity and fear of crime', *Environment and Behaviour* 17(3):275–95.

London Rape Crisis Centre (1984) *Sexual Violence*, London, Women's Press.

Matrix (eds) (1984) *Making Space: Women and the Man-made Environment*, London, Pluto Press.

Moore, C. and Brown, J. (1981) *Community versus Crime*, London, Bedford Square Press.

Morley, D. (1986) *Family Television: Cultural Power and Domestic Leisure*, London, Comedia Publishing Group no. 37.

Perkins, D. E. (1986) 'Sex offending: a psychological approach', in J. Hollins and K. Howells, *Issues in Criminological and Legal Psychology* 9, Leicester, British Psychology Society.

Riger, S. and Lavrakas, P. (1981) 'Community ties: patterns of attachment and social interaction in urban neighbourhoods', *American Journal of Community Psychology* 9:653–65.

Sayers, J. (1982) *Biological Politics: Feminist and Antifeminist Perspectives*, London, Tavistock Publications.

Souster, M. (1989) 'Bad lighting link reinforced', *Times* 9 March.

Southwark Borough Council (1984) *A Safe Place to Live*, London, Southwark Borough Council.

Stanko, E. A. (1987) 'Typical violence, normal precaution: men, women and interpersonal violence in England, Wales, Scotland and USA', in J. Hanmer and M. Maynard (eds) *Women, Violence and Social Control*, Basingstoke, Macmillan, 122–34.

Toseland, R. W. (1982) 'Fear of crime: who is most vulnerable', *Journal of Criminal Justice* 10:199–209.

Valentine, G. M. (1989a) 'Women's fear of male violence in public space: a spatial expression of patriarchy', Ph.D. thesis, Department of Geography, University of Reading.

—— (1989b) 'The geography of women's fear', *Area* 21(4):385–90.

—— (1990) 'Women's fear and the design of public space', *Built Environment* 16(4): 288–303.

van Dijk, J. J. M. (1978) 'Public attitudes toward crime in the Netherlands', *Victimology* 3(3/4):265–73.

Wyre, R. (1986) *Women, Men and Rape*, Oxford, Perry Publications.

8

RACE AND ETHNICITY

Malcolm Cross

I

It is now widely understood that London's crisis is one of growing spatial and social division. Studies of the richest and poorest parts of the city have shown that, even when employment chances were rising overall, the gulf has widened over the last two decades (Congdon 1989). In this chapter I want to argue three things: first, that these emerging splits are *both* spatial and social but that the causes differ; second, that race and migration are intimately bound up with these changes; and, third, that this is true in ways which are rarely understood. The point of the last is to suggest that what we are witnessing is a new division of work and worklessness which impacts on ethnic minority groups in ways which divide them from each other, as well as from the majority society. In this sense, therefore, the crisis of London is intimately entwined with issues of race, but not in a simple sense. The city sorts and filters; there is more than one 'front line'.

London is, of course, changing rapidly. As a 'global city' it has been at the forefront of economic restructuring, as we saw in Chapter 3. Manufacturing, other production industries and transport/wholesaling jobs fell by 34.6 per cent between 1971 and 1988, with a corresponding rise in the service sector (Gordon *et al.* 1991). Within the region, the decentralization process continued with total jobs in Greater London falling by 430,000 and rising by 370,000 in the Outer Metropolitan Area over the same period. A feminization process is also evident, as is the decline in unskilled jobs.

Spatial concentration

There seems little reason to doubt that blacks and Asians are still heavily concentrated geographically. Angus Stuart (1989), for example, calculates, using the Longitudinal Sample from the 1971 and 1981 Census, that 60 per cent of Asians and nearly three-quarters of Afro-Caribbeans live in the four main conurbations.

In Greater London the distribution of ethnic minorities is far from even. As Figures 8.1 and 8.2 show very clearly, there are very few places in the city where Afro-Caribbean and Asian concentrations occur simultaneously. What this

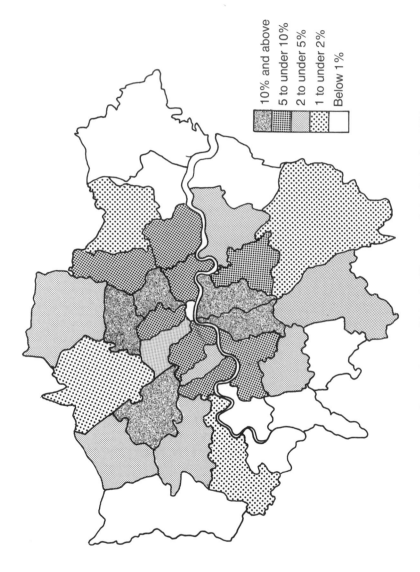

10% and above

5 to under 10%

2 to under 5%

1 to under 2%

Below 1%

Figure 8.1 Distribution of Afro-Caribbean population in London 1981

Source: London Research Centre.

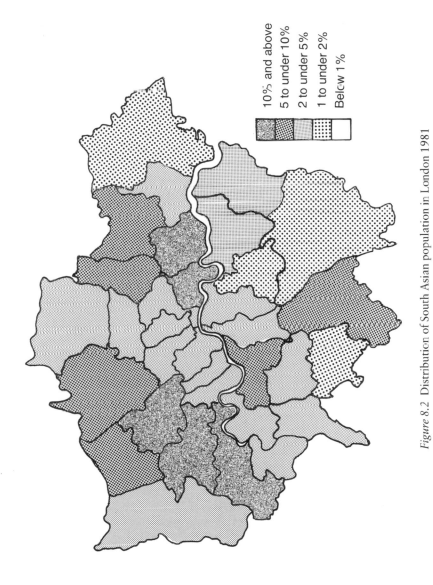

10% and above

5 to under 10%

2 to under 5%

1 to under 2%

Below 1%

Figure 8.2 Distribution of South Asian population in London 1981

Source: London Research Centre.

means in practice is that, notwithstanding the communities of poor Bengalis in Tower Hamlets (Eade 1989), the Asian population is not over-represented in the so called 'inner city'. Afro-Caribbean and Asian groups are under-represented in the central business district, but only Afro-Caribbeans are concentrated in the inner city, where they are 80 per cent more likely than whites to live.

Afro-Caribbeans are also less likely to migrate internally (Robinson 1991). The pattern for Greater London confirms this national picture. What is evident is that only a small proportion of Caribbean-British people left London over the period 1971–81 when compared with others (Census). More than a fifth of whites had left London at the end of the decade compared with only one in ten from minorities, but Afro-Caribbeans were both more likely than other groups to remain where they were and also this was far more likely to be the 'inner city'. Obviously, this is partly a function of age, since more of the whites would be retiring to the countryside, but this is not the complete picture. Afro-Caribbeans, who are already concentrated in Inner London, are more likely than others to stay there. Four out of five Afro-Caribbeans stayed put, while – at the other extreme – almost half the 'inner city' Indians moved out, most of them to the suburbs.

What is particularly important, then, is that Afro-Caribbeans and Asians occupy clearly different spatial locations in the city's social structure. By and large the decline in employment in Greater London has been offset by a decline in labour supply as more people move out to the suburbs and beyond. This has not been the case with Afro-Caribbean labour. To some extent, therefore, the Asian population, with the probable exception of the Bengali population in East London, shows a pattern of suburbanization more like the majority population.

In purely spatial terms, therefore, ethnic minorities are not all in the same position in Greater London. This is not simply true in settlement terms but also dynamically; that is, over time the differences are widening as London comes to have two areas of ethnic minority concentration: the 'inner city' boroughs of Hackney, Haringey, Lambeth, Lewisham and Wandsworth and the 'outer city' boroughs of Brent, Ealing, Hounslow and Waltham Forest. These are strongly and increasingly correlated with different groups. The existence of counter cases (e.g. Tower Hamlets and Brent) does more to illustrate the point than to confound it.

Social polarization

There is little doubt that ethnic minorities across the country as a whole are concentrated in industrial sectors of major decline. For males of Afro-Caribbean origin, for example, there is a concentration in the sectors of mechanical engineering, vehicle repair and transport, reflecting the maintenance of patterns laid down in the early years of migration and reproduced among the children of settlers (Brown 1984). Asian men are also unlikely to be found in the 'primary' industrial sector (agriculture, mining, etc.) and to be over-represented in both non-metal manufacturing industries, metal manufacture (Indian only) and in 'distribution, hotels and catering, repairs'. All those of New Commonwealth (or Pakistan) origin and their descendants are under-represented in the growing sectors of banking, finance and other services. The amalgamated rounds of the Labour Force Survey for 1985–7 show that a shift has occurred out of metal manufacture into other forms of manufacturing. Moreover, there has been a shift into services. For white men this now accounts for 27 per cent of the total, while for ethnic minorities it has risen to 21 per cent (Department of Employment 1988b).

Among women the concentrations are equally clear but perhaps less precise: there is a dramatic over-representation of Afro-Caribbean women in service jobs. There was also a slight over-representation of Asian (Indian) and Afro-Caribbean women in the transport sector. By 1985–7, the concentration of largely Asian women in non-metal manufacturing was still evident, together with a growing concentration in the service sector. For example, 52 per cent of white women workers are in the service sector, compared with 47 per cent of minority women.

The pattern in Greater London is different. In Peter Townsend's survey of 1986 (Townsend et al. 1987), for example, the over-representation of Afro-Caribbeans in manufacturing was not then evident. The Longitudinal Study from the 1971 and 1981 Censuses shows that Afro-Caribbean people in Greater London had come by 1981 to be only slightly over-represented in manufacturing, whereas other Asian groups were still strongly maintaining a concentration in this field. Both Asian men and women have also penetrated the banking and insurance sectors. For Afro-Caribbeans of either gender the emerging concentration is in the secondary services sector and there is some evidence that they are coming to have a similar representation to the population at large employed in the public sector.

Overall in Greater London, the decline over this decade was indeed in lower-level operative and labouring posts. Women's jobs rose by an eighth while men's fell by a fifth. As far as minorities are concerned, both ethnic group and gender were significant, particularly the latter. Afro-Caribbeans tended disproportionately to move into lower-level services while Indian men increased their representation relative to whites in managerial positions. Gender, however, played a more important role. Ethnic minority women when compared with whites moved out of unskilled operative posts and into white-collar secretarial jobs.

What is perhaps of greater importance is that there are marked disparities in the socio-economic position of ethnic minorities in Greater London. This can be seen in Figure 8.3 which uses an index comparing the representation of minorities with that of the white population in different socio-economic categories. What it shows is that Afro-Caribbeans are indeed markedly under-represented at the top end of the class structure of London while for 'South Asians' their distribution is bimodal. They are strongly over-represented at the top end but also in semi-skilled employment.

Evidence on social mobility, or change in socio-economic profiles over time, is hard to come by. What we have comes from the same data sources used here

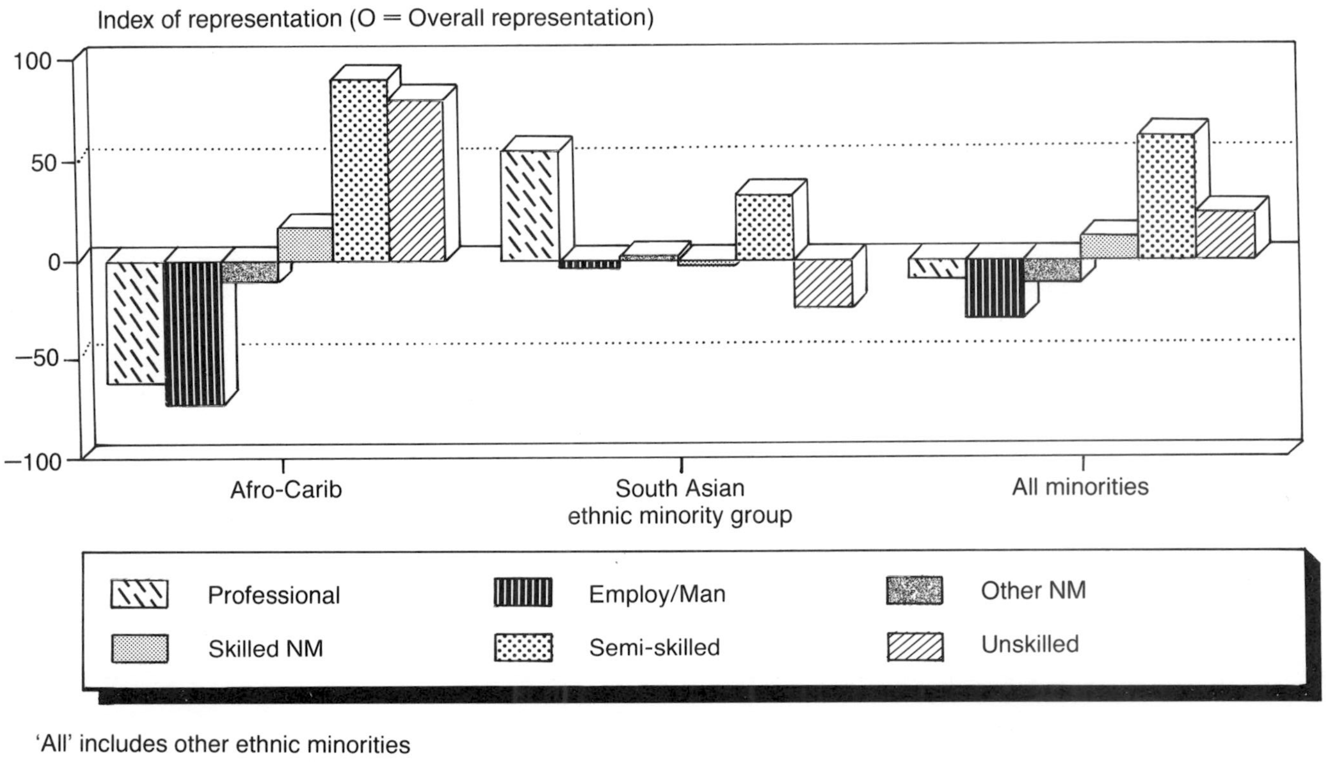

Index of representation (O = Overall representation)

Figure 8.3 Greater London SEG representation by ethnic group 1981

Source: London Research Centre / DT5741.

'All' includes other ethnic minorities

Professional

Skilled NM

Employ/Man

Semi-skilled

Other NM

Unskilled

Afro-Carib

South Asian
ethnic minority group

All minorities

100

50

0

−50

−100

(Robinson 1990) and tends to show nationally that for the Afro-Caribbean population their under-representation at higher socio-economic group (SEG) levels has remained more or less constant, while for Asians there are some signs of upward movement. For Greater London, the picture depends very clearly on which groups are being compared with the population of the city as a whole. For example, over the intercensal decade 1971–81, Indians came to be over-represented in 'higher white-collar' jobs (professional/managerial). There was some improvement from a very low level for Afro-Caribbeans and almost no change at all for those of Pakistani or Bangladeshi origin.

The evidence has been mounting for some time that ethnic minorities, with their origins in recent waves of migrant labour, are deeply affected in employment terms both by recession and by restructuring. From a position of relatively high levels of employment in the early years they are now being forced disproportionately into the ranks of the workless (Ward and Cross 1990). With continuing high rates of labour market participation and age profiles skewed towards the young, they are likely to be major victims of recession and restructuring wherever it occurs. The Labour Force Survey also shows that the gap between white and black unemployment rates is at least as wide at higher skill levels as at lower. For example, the ratio between white and black unemployment levels for men 16–64 is 1:1.8, but for those with higher level qualifications it is 1:3 (DoE 1988b: 642).

In London, the evidence suggests that the loss of labour from declining sectors has been achieved by a decline in recruitment rather than by redundancies or other forms of departure. This has disproportionate effects on populations with skewed age distributions towards the young. If we control for age group, however, as in Figure 8.4 the pattern that emerges is more complex. The Afro-Caribbean population is indeed at least twice as likely to be unemployed at each age group as the white population. This is far less true for Asians, where it is the young who are vulnerable together with older groups of Pakistani and Bangladeshi workers for whom language barriers may be an additional factor.

A key question to which the foregoing gives rise is whether the spatial distribution of ethnic minorities across London accounts for their socio-economic position. Is London's Caribbean-British population so poorly placed in social class terms *because* it is concentrated in the 'inner city' where jobs are few and largely unskilled? To some extent this is inevitably so: the more one group is concentrated in an area of limited opportunity the more that group as a whole will suffer from the constraints of those districts. What does not follow, however, is that their position *relative to whites* will be influenced by where they live. Blacks tend to be *more* over-represented at lower socio-economic levels *outside* the inner city than within. In other words, 'race' appears to make more of a negative difference in areas of the city where black people are less likely to be found. Again, this is not true for South Asians. For them the bimodal distribution holds true for both the inner city and suburbs; it is only in the Central Business District itself that this is not true. Here, the Asian population is more professional and middle-class than the population as a whole.

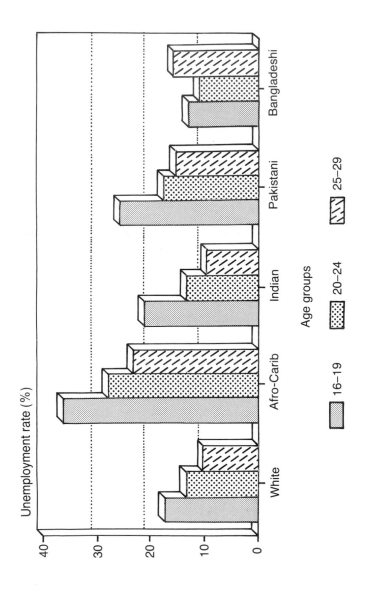

Figure 8.4 Unemployment rates by race and age, Greater London 1981

Source: Census 1981.

II

In the second half of this chapter, I want to ask where this evidence takes us in determining the importance of race and ethnicity in understanding the way that London is changing. I shall look first at studies which are specifically concerned with racial minorities, then at approaches to changing labour markets in urban contexts, and finally at what I shall call 'ethnic division of labour theory'.

Race and racialization

It could be said that recent debates have not selected for serious attention the issues with which this chapter is concerned. For example, in the long-running 'race' versus class debate on the origins and evolution of racism, there has been very little concern with either spatial or labour market issues. Older theories of racial formation, which sought to identify the field for specialist attention on the basis of one interpretation of Weberian theory, made passing reference to occupational and industrial distributions, but eschewed the analysis of change (Rex and Tomlinson 1979). Critics of this view, while often interested in change processes, have been more preoccupied with arguing for the subordinate status of the 'racialization' of the workforce to the incorporation of migrant labour at a particular phase of capitalist development (Miles 1982; 1986; 1987). This perspective on the analysis of capitalism tends to flatten the terrain of individual histories, let alone specific cities or different minority groups within them.

The recent book by Susan Smith (1989) is an exception to this tradition. The main thrust of the book is to argue that patterns of residential segregation have been generated in part at least by political acts, sustained by 'common-sense racism':

> As a residential pattern ... specifically *racial* segregation reflects and structures enduring inequalities in access to employment opportunities, wealth, services and amenities, and to the package of civil and political rights associated with citizenship. As an ideology, on the other hand, segregationism builds from the objective deprivation of black people to a subjective acceptance that racial differentiation has a logic of its own; it provides a reservoir of common sense justifications for discriminatory policy and separatist practice.
>
> (Smith 1989: 170)

This is an important contribution precisely because it brings space back onto the agenda for the study of racialized inequality. What it does not do is point out how the social organization of space is changing in response to newly emergent patterns of employment. More particularly, this study specifically elides all non-white ethnic groups. As I have shown such aggregation is totally untenable in the London case. What studies of ethnic minorities have shown beyond reasonable doubt is that discrimination in housing, employment and in many other spheres of living is an everyday reality for Afro-Caribbean, Asian and other identifiable minorities in London as elsewhere in the UK. What they have not shown is the extent to which minorities are structurally located at different points of the city. Whatever discrimination may do, therefore, it does not produce uniformity of outcome. Moreover, nearly all analyses of ethnic minorities tend to assume that racial exclusion is itself an undifferentiated phenomenon, rather than a selective process. Before returning to this theme, we need to consider whether theories of labour market change are themselves more helpful.

Labour market change

Theories of labour market change, as distinct from theories on the structure of the labour market itself, are nearly all premised upon posited changes in the nature of capitalist (and possibly non-capitalist, industrial) economies. These changes are those most often associated with so called 'post-industrialism', 'post-Fordism' and even 'postmodernism' (Albertson 1988; Cooke 1988; Jameson 1984). They include the decline of manufacturing, the growth of services and the spatial reorganization of production (Buck 1985). The increase in the service sector has been led by public sector welfare posts and, more recently, by growth of business services, including finance and insurance (Martin 1988). These changes have been accompanied by the increasing feminization and casualization of the labour force and the emergence of a new social division of labour (Pahl 1984: 1988).

It is very striking that while political and popular opinion appears to accept the salience of race in economic restructuring, the most important analyses from the research community have been muted on this theme (Buck *et al.* 1986; Spencer *et al.* 1986). They have often *assumed* an implication for minorities without exploring whether those who are differentially located are affected in the same way. Still less have they explored the potentially recursive aspects of the phenomenon: namely whether the emergence of racialized groups contributes itself to the processes of spatial and social sorting. I shall not consider this important point here but simply look at two lines of theoretical development which we might call 'spatial polarization theory' and 'social polarization theory' without wanting to suggest that these are always easy to differentiate in practice. It is striking, for example, that – as far as race is concerned – the delineation of the first is largely American (where it is assumed to encompass the second). The second is largely home-grown, where, if anything, it is assumed to embrace the first.

Spatial polarization theory

A major distinction in labour market theory is that between the thesis of discontinuity and continuity. Normally, this is considered a distinction between classical economic models and those of a more sociological persuasion, with the latter positing the existence of major splits in the structure and operation of the labour market. When applied to labour market change, this distinction has tended to emerge in the debate between 'mismatch theory' and 'social polarization' perspectives.

The former view starts by noting the indisputable shifts in the demand for labour. John Kasarda, for example, argues that similar processes of industrial transformation have occurred in US and European cities, leading to a decline in the opportunities they offer for occupational mobility. When compared with the past, the employment gains these cities have shown have been predominantly in higher-level positions demanding educational and technical qualifications (Kasarda *et al.* 1992).

These changes are seen as occurring simultaneously with another, and possibly older, process. This is the selective outmigration of urban residents, in which white, middle-class families move out, initially to the suburbs and later to the small towns to become a new rural bourgeoisie. The result is the 'mismatch' between the supply of appropriately qualified labour and the newly rising demand for technical qualifications.

This supply-side approach is, however, only one variant of spatial polarization theory. Another concentrates upon the failure of demand in certain zones to keep pace with labour supply. The diagnosis is similar, though the aetiology varies. The argument is not that particular groups have failed to utilize opportunities for advancement but that previous experience of social exclusion, coupled with the new spatial division of labour, creates conditions in which an 'underclass' emerges which in time does indeed come to possess characteristics of a pathological kind. These might include a high incidence of family breakdown, drug abuse and crime. The most important statement of this position is by William J. Wilson, who argues that in many US cities spatial decline and the emergence of an increasing black poor have led to social isolation. As he writes:

> the communities of the underclass are plagued by massive joblessness, flagrant and open lawlessness, and low-achieving schools and therefore tend to be avoided by outsiders. Consequently, the residents of these areas, whether women and children of welfare families or aggressive street criminals, have become increasingly socially isolated from mainstream patterns of behaviour.
>
> (Wilson 1987: 58)

An important point to note is that evidence against this phenomenon cannot be derived from figures showing the distribution of blacks across the class structure (Gallie 1988). Indeed Wilson's thesis is that anti-discrimination and equal opportunity measures have helped to create a growing black middle class. The existence of this group compounds rather than relieves the problem, since newly socially mobile blacks follow whites to the suburbs and thereby reduce the pool from which effective community leaders might be drawn. In periods of resurgent affluence, therefore, the increasingly black poor remain in conditions of mounting unemployment. They become, in Wilson's depressing phrase, the 'truly disadvantaged' (1987).

The major difference between these two variants of spatial theory lies in their policy prescriptions. The former view, concentrating as it does on labour quality, tends to direct attention away from welfare support – since that can be regarded as locking the poor into those areas where their prospects are least – and towards encouraging outmigration and the gentrification of cities. The second takes the opposite line in demanding more schools, better training, and welfare support as the only way to close the gap and reduce the 'social isolation' of inner city disadvantage.

Social polarization theory

If theories of spatial polarization are premised upon a belief in the emergence of areas of joblessness and social dislocation, social polarization theory owes its origins much more to class theory, that is to a relational approach. As a development of segmentation and dualism, social polarization perspectives start from a structural conception of discontinuity in the demand for labour power. Capitalist economies are passing through a phase of ever widening divisions between an enhanced level of demand for higher-level positions, consequent upon the growth of finance capital, coupled with a much lower level of demand for low-cost, flexible, non-unionized (and often female) labour power. Thus, contrary to spatial theory, it is not the lowest-level jobs that are missing or declining but those in between. This implies a decline in social and geographical mobility options, particularly for ethnic minorities, who remain available only for inclusion in the lower sector of the market. The key indicators of social polarization lie in income distributions, rather than in social problem indices (Buck 1991).

It would be wrong to suppose that this approach has no specific implications for urban labour markets. The operations of global capitalism may generate global cities whose fortunes depend on the interplay of finance capital and global trade. The work of Saskia Sassen (1988) is an excellent example. Major international cities, sited at the nodal points for the control of capital flows in the global economy, experience a rapidly growing demand for the servants of finance capital while at the same time the demand for low-level services (cleaning, catering, etc.) also grows apace. These demands may be met internally, particularly where female labour power is available. More probably, however, they will be met externally by the renewed import of migrants. This new migration will be even more short-term, insecure and exploited than before, since many of the legal, health and safety measures will have been repealed to facilitate capital flows. In particular, the new demand is likely to help sustain refugee flows and undocumented migration.

The ethnic division of labour

From the point of view of charting the fortunes of ethnic minorities the differences between these two perspectives are not clear-cut. Both theories predict a sustained downturn in minority fortunes, either because of lower levels of educational attainment and high levels of concentration in areas of industrial decline or because middle-sector jobs – which might have provided stepping-stones for social mobility – have disappeared. If the new service needs of the global city are filled by a 'new helotry' (Cohen 1987) then the prospect may be a choice between engaging in the labour market at that level or sitting it out in poverty on the sidelines.

A third view, which is particularly geared towards explicating the complex encounter between minorities and changing labour markets, is beginning to emerge. It is a view which suggests that, notwithstanding high levels of racism, ethnic minorities may be affected differently by these changes. Bailey and Waldinger (1988), for example, writing on the fortunes of Hispanics, Black Americans and 'Asians' (Far Eastern origin) in New York between 1970 and 1986, show that blacks were sheltered in the early part of this period by being disproportionately located in the public sector, rather than in manufacturing, but this shelter later kept them out of the sun as other areas took off in employment terms. Moreover, an increasing proportion of blacks were falling by the wayside in remaining among the long-term unemployed.

Hispanics in New York, by contrast, show a very different pattern. Although heavily dependent on a declining manufacturing base, their position in this sector actually strengthens over the period because the white exodus is even more evident than the decline in jobs overall. Waldinger and Bailey cite the garment industry as one obvious example. Foreign-born Asians experienced the transitions in the labour market in a yet different fashion. Although originally concentrated in manufacturing and retail, this group was well placed to exploit the rapid rise in financial and related services.

Ethnic division of labour theory is important precisely because it does not fall into the trap of accounting for differences in terms of cultural traditions or of denying their existence because of an adherence to blanket racism theory. It is the *structural* position of minorities which determines their experience of labour market change. These positions are not only influenced by labour demand changes but also by the relative size (and change) of minorities, their particular concentration in certain sectors and industries and their capacity to shift from one sector to another.

Conclusion

Each of the theoretical strands of thought outlined isolates a part of the reality. Spatial polarization theory is helpful in describing the areal effects of post-industrial change. It is clearly true that minorities are highly vulnerable to these changes. Their vulnerability does not, however, spring from their lack of adaptation to job demands. It comes more from an areal concentration which may be likely to impact differentially on Afro-Caribbean populations. The division in the labour market does not itself show conclusive evidence of dualism, since ethnic minorities are distributed almost across whole spectra of occupational positions. It may come from discrimination effects which constrain achievement without denying differentiation. This differentiation is important since interactions between race, class and place are not themselves uniform and dualistic. Afro-Caribbean men, and possibly men and women from some Asian origins, are the worst affected. For them vulnerability exacerbated by labour market change is a compound problem.

Although structural position may differ, there is also the possibility that these differences have been mediated by different strategies of exclusion on the part of the majority population. In the conventional language of this discourse, racism would then not be regarded as a unitary phenomenon but rather one with at least two faces. The first is the colonial tradition of biological racism which channels those so stigmatized into unemployment, run-down areas and confrontations with agencies of social control. The second is the exclusion of the culturally different who are subject to abuse, attack and canalizing into unwanted occupational niches. Whereas the former helps to generate the spatial ghetto, the latter produces the occupational ghetto (Cross and Johnson 1988). This approach has recently figured in French debates, where the distinction is between the 'external' threat of 'Third Worldization', as distinct from the 'internal' exclusion of – in this case – the black French people (Antilleans) (Balibar and Wallerstein 1988). Clearly, if either structural location in the labour market before its post-industrial transformation or independent processes of exclusion are operative, then we should not expect to see one set of outcomes but rather effects which are group-specific.

'Race' is therefore integral to the crisis of London but in two quite separate ways. First, it is coming increasingly to define an urban underclass, marooned in crumbling estates and weighed down by poverty, joblessness and despair. Second, however, it is a rising marker of enforced occupational specialization in which cohorts of young people from largely Asian backgrounds are joined by new migrants to become the 'new helots', or those on whom the transition to global city depends. Gender divisions cut across this generalization. Afro-Caribbean women, for example, are also incorporated as the lower part of the new service class. Their wages are unlikely to prove sufficient, however, to prevent ghettoization from scarring the face of the city.

REFERENCES

Albertson, N. (1988) 'Postmodernism, post-Fordism and critical social theory,' *Environment and Planning D: Society and Space* 6:339–65.
Bailey, T. and Waldinger, R. (1988) 'Economic change and the ethnic division of labor in New York City', paper prepared for the Social Science Research Council, Committee on New York City, Dual City Project.

Balibar, E. and Wallerstein, I. (1988) *Race, Nation, Classe; les identités ambigues*, Paris, La Decouverte.

Body-Gendrot, S. (1989) 'Migration and the racialisation of urban space in France', paper presented to the conference on 'Racism and the postmodern city', University of Warwick, 28–31 March.

Brown, C. (1984) *Black and White Britain*, London, Heinemann.

Buck, N. (1985) 'Service industries and local labour markets: towards an anatomy of service employment', paper presented to the Regional Science Association Annual Conference, September.

—— (1991) 'Social polarisation in the inner city: an analysis of the impact of labour market and household change', in M. Cross and G. Payne (eds) *Work and the Enterprise Culture*, London, Falmer Press.

Buck, N., Gordon, I. and Young, K. (1986) *The London Employment Problem*, Oxford, Oxford University Press.

Cohen, R. (1987) *The New Helots: Migrants in the International Division of Labour*, Aldershot, Avebury.

Cohen, R. B. (1981) 'The new international division of labor, multi-national corporations and urban hierarchy', in M. Dear and A. J. Scott (eds) *Urbanisation and Urban Planning in Capitalist Society*, London/New York, Methuen, 287–315.

Congdon, (1989) 'An analysis of population and social change in London wards in the 1980s', *Transactions of the Institute of British Geographers* N.S. 14: 478–91.

Cooke, P. (1988) 'Modernity, postmodernity and the city', *Theory, Culture and Society* 5:475–92.

Cross, M. (ed.) (1992) *Ethnic Minorities and Industrial Change in Europe and North America*, Cambridge, Cambridge University Press.

Cross, M. and Johnson, M. R. D. (1988) 'Mobility denied: Afro-Caribbean labour and the British economy', in M. Cross and H. Entzinger (eds) *Lost Illusions: Caribbean Minorities in Britain and the Netherlands*, London, Routledge, 73–105.

—— (1989) *Race and the Urban System*, Cambridge, Cambridge University Press.

Department of Employment (1988a) 'Ethnic origins and the labour market', *Employment Gazette* March: 164–77.

—— (1988b) 'Ethnic origins and the labour market', *Employment Gazette* December: 633–46.

Eade, J. (1989) *The Politics of Community: The Bangladeshi Community in East London*, Aldershot, Gower.

Gallie, D. (1988) 'Employment, unemployment and social stratification', in D. Gallie (ed.) *Employment in Britain*, Oxford, Blackwell, 465–92.

Gilroy, P. (1987) *There Ain't no Black in the Union Jack*, London, Hutchinson.

Gordon, I., Fainstein, S. and Harloe, M. (eds) (1991) *A Tale of Dual Cities: London and New York*, Oxford, Basil Blackwell.

Henderson, J. and Castells, M. (eds) (1987) *Global Restructuring and Territorial Development*, London, Sage.

Jameson, F. (1984) 'Postmodernism, or the cultural logic of late capitalism', *New Left Review* 146:53–92.

Kantor, P. and Stephen, D. (1988) *The Dependent City: The Changing Political Economy of Urban America*, New York, Scott Foresman.

Kasarda, J. D. (1988) 'Jobs, migration and emerging urban mismatches', in M. G. H. McGeary and L. Lynn (eds) *Urban Change and Poverty*, Washington, DC, National Academy Press.

Kasarda, J. D., Friedrichs, J. and Ehlers, K. (1992) 'Urban industrial restructuring in the US and West Germany', in M. Cross (ed.) *Ethnic Minorities and Industrial Change in Europe and North America*, Cambridge, Cambridge University Press.

117

Lash, S. and Urry, J. (1987) *The End of Organised Capitalism*, Cambridge, Polity Press.

Martin, R. (1988) 'Industrial capitalism in transition: the contemporary reorganisation of the British space economy', in D. Massey and J. Allen (eds) *Uneven Redevelopment: Cities and Regions in Transition*, London, Hodder & Stoughton.

Miles, R. (1982) *Racism and Migrant Labour*, London, Routeldge & Kegan Paul.

—— (1986) 'Labour migration, racism and capital accumulation in Western Europe since 1945: an overview', *Capital and Class* 28:49–86.

—— (1987) *Capitalism and Unfree Labour: Anomaly or Necessity*, London, Tavistock.

Noyelle, T. and Stanback, T. (1984) *The Economic Transformation of American Cities*, Totowa, NJ, Rowman & Allenheld.

Omi, M. and Winant, H. (1986) *Racial Formation in the United States*, London and New York, Routledge & Kegan Paul.

Pahl, R. E. (1984) *Divisions of Labour*, Oxford, Basil Blackwell.

—— (ed.) (1988) *On Work: Historical, Comparative and Theoretical Approaches*, Oxford, Basil Blackwell.

Rex, J. A., and Tomlinson, S. (1979) *Colonial Immigrants in a British City: A Class Analysis*, London, Routledge & Kegan Paul.

Robinson, V. (1990) 'Roots to mobility: the social mobility of Britain's black population, 1971–87', *Ethnic and Racial Studies* 13:274–86.

—— (1991) 'Goodbye yellow brick road: the spatial mobility and immobility of Britain's ethnic population, 1971–1981', *New Community* 17(3):313–30.

Sassen, S. (1988) *The Mobility of Labor and Capital*, Cambridge, Cambridge University Press.

Sassen-Koob, S. (1984) 'The new labor demand in global cities', in M. P. Smith (ed.) *Cities in Transformation: Class, Capital and the State*, Urban Affairs Annual Review vol. 26, Beverly Hills, CA, Sage, 139–71.

Sawers, L. and Tabb, W. K. (1983) *Sunbelt/Snowbelt: Urban Development and Regional Restructuring*, Oxford, Oxford University Press.

Smith, M. P. and Feagin, J. (eds) (1987) *The Capitalist City: Global Restructuring and Community Politics*, Oxford, Blackwell.

Smith, S. J. (1989) *The Politics of 'Race' and Residence: Citizenship, Segregation and White Supremacy in Britain*, Cambridge, Polity Press.

Spencer, K., Taylor, A., Smith, B., Mawson, J., Flynn, N. and Batley, R. (1986) *Crisis in the Industrial Heartland: A Study of the West Midlands*, Oxford, Oxford University Press.

Sternlieb, G. and Hughes, J. (1976) *Post-Industrial America: Metropolitan Decline and Inter-Regional Job Shifts*, New Brunswick, NJ, Centre for Urban Policy Research.

Stuart, A. (1989) *The Social and Geographical Mobility of South Asians and Caribbeans in Middle-Age and Later Working Life*, LS Working Paper no. 61, London, City University.

Thrift, N. and Williams, P. (eds) (1987) *Class and Space*, London, Macmillan.

Townsend, P., Corrigan, P. and Kowarick, V. (1987) *Poverty and Labour in London*, London, Low Pay Unit.

Waldinger, R. (1988) 'Immigration and urban change', unpublished paper.

Ward, R. and Cross, M. (1990) 'Race, employment and economic change', in P. Brown and R. Scase (eds) *Poor Work: Disadvantage and the Division of Labour*, Milton Keynes, The Open University Press, 116–31.

Wilson, W. J. (1987) *The Truly Disadvantaged: The Inner City, the Underclass and Public Policy*, Chicago, University of Chicago Press.

9

THE REBIRTH OF COMMUNITY PLANNING

George Nicholson

The city is a people's art. True involvement comes when the community and the designer turn the process of planning and building a city into a work of art.

(Edward Bacon 1967)

INTRODUCTION: WHAT KIND OF CITY DO WE WANT?

Each city is unique. The cultures, functions and history which collectively give to a city its own individual weave are products of a special set of circumstances. Often this simple fact is ignored or even derided as the city finds its way onto the drawing boards of planners, architects or developers. The inevitable result is that many of the decisions, plans and development proposals that have emerged in London and other cities have been decidedly 'anti-city' both in their conception and their impact. This is what makes the question 'What sort of city do we want?' such an important one, yet today it is often left unasked. Our cities will remain battle-grounds until we address this question. London is a good example; a historic city which has experienced a number of cycles each of which has left its mark. Could it have been different? This question has to be framed within the historical realities of a country's particular patterns of land ownership and political hegemony. However, there are undoubtedly lessons to be learnt from the history of urban development in London. Two almost forgotten, and long overdue for revival, can be found inside the 1944 Greater London Plan, a historic landmark in planning in the UK. In it Sir Patrick Abercrombie proposed two important principles, both as relevant today as they were then. One was that planning is a matter of 'grafting new growth onto the old stock of London'. If that principle alone had been adhered to, much of the damage to the historic fabric of many great cities throughout the world could have been avoided. The other principle was equally significant. It was that the community structure of London should be regarded as 'the basic planning unit' (see Figure 9.1). This was developed at length in a chapter titled 'Community planning'. Indeed Abercrombie held that 'the community idea' should dominate both the 1943 County of London Plan and the 1944 Greater London Plan (Abercrombie 1945).

119

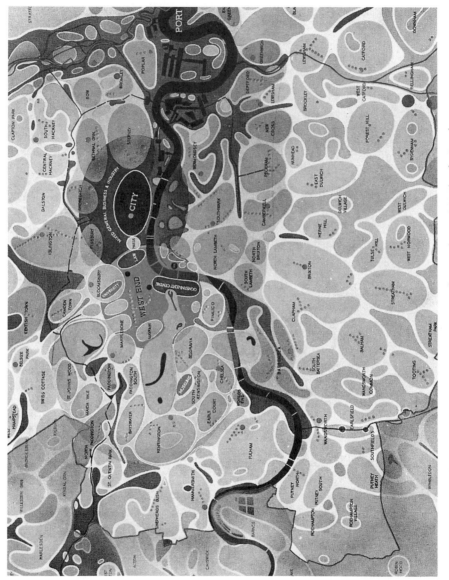

Figure 9.1 Community structure in Abercrombie's 1944 Plan for London

Here then is a clear conception, a starting-place, a purposefulness on which to build the future of a city.

TRUSTING THE PEOPLE!

The situation facing communities in Central London for the last four decades has been one of being squeezed from a number of directions. This stems from a mixture of the side-effects of past planning policies, intense development pressures emanating from the central commercial core, government policy, and local authority investment decisions.

Planning policy from the last war up to the late 1970s encouraged decentralization of homes and jobs from Central London. A study of the resident population of Central London by community organizations in 1980 was instrumental in highlighting that one in every four residents had been lost in just ten years – a rate of decline more than five times as fast as that in Outer London. Over the same period, Inner London also lost 41 per cent of its manufacturing jobs. Some commentators point out that many Londoners during this period were simply voting with their feet, rather than moving as a result of a deliberate policy. Either way, what resulted was the resident and working base of Central London being dramatically reduced.

With proper monitoring of what was happening, policy changes might have been formulated sooner than they were. However, the situation in London in the 1970s was that not only was little monitoring being done of the effects of past planning policies, but the master plan containing London's future planning policies, the Greater London Development Plan (GLDP) was already out of date when it was published in 1976. This was because central government sat on the document for 7 years after it had been submitted by the Greater London Council. The fact that this is still the master plan for London, 16 years later, gives some idea of the scale of the present malaise affecting policy formulation and decision-making in London. The GLC, shortly before it was abolished, did update the GLDP, but the government refused to adopt the plan. Instead the same legislation which dispatched the Greater London Council contained proposals to scrap the GLDP altogether, replacing it with government-provided 'strategic guidance'. Monitoring by the London Planning Advisory Committee (LPAC) of the emerging new-style local plans that the government has also put in place – the Unitary Development Plans – confirms the concerns expressed during the abolition debate about greater incoherence in planning the capital after the demise of the GLC. London decision-makers have thus been ill-prepared and ill-equipped over a prolonged period.

The situation would have been bad enough in managing the legacy of past planning policies, and in dealing with normal development pressures. But it proved even more of a handicap in coping with the side effects of the commercial property boom which Central London faced in the 1970s and 1980s. Such booms compound the situation in a number of ways. By driving up land prices, they encourage speculation in land and building. This leads to plant closure and loss of industrial land. London's Docklands is a classic example of this process in action (Docklands Consultative Committee 1990). In addition house prices are pushed beyond the means of local people and land for social housing becomes prohibitively expensive. Housing land therefore becomes very short in supply by virtue of both cost and quantity. The same applies to industrial land. Speculation also leads to loss of local shopping and other community facilities. The tendency of this process is towards a monopoly of high-value uses and a consequent under-mining of the previous mix of uses and range of opportunities.

Two further factors have an important bearing. First, improvement, modern-ization and redevelopment of social housing has been a major contributor to the decline in population in Central London, although this process is now almost complete. In addition the effects of gentrification and the increasing number of second homes has also led to a decline in population. Second, the decline in public investment in housing and the shrinkage in the private rented sector has led to a decline in the supply of affordable homes. This is likely to be made worse by the government's recent housing legislation, as described in Chapter 2.

As the Campaign for Homes in Central London (CHiCL) report 'Wanted – dead or alive' (submitted to the House of Commons Environment Committee in March 1983) put it:

> The dramatic loss of resident population, especially of households with children, has undermined community life, and now threatens the viability of central London's last remaining residential communities. It has:
>
> led to the closure of essential shops and schools. Central London is being pulled into a vicious circle of decline, where communities can no longer support new or existing residents.
> led to an imbalance in the remaining population.
> led to an increasing shortage of essential service workers.

Since the report the problems have accelerated owing to the combined effects of recent changes in planning and housing policy, and a property boom this time embracing the residential market as well as the commercial property sector. The combined effects of government legislation concerning council house sales, public expenditure, deregulation of planning and housing, GLC abolition and changes in fiscal policy have created the conditions for even more intense pressures on Central London communities than those of the 1970s.

Whereas the concerns being expressed by community organizations had been pretty much a lone voice in the 1970s and early 1980s, the dramatic rise in house prices in the late 1980s has led to widespread concern at the diminishing supply of affordable housing and the consequent distortions being caused in the labour market by staff shortages.

The recent study by CHiCL, *Access to Housing: the Problems for Employers and Potential Employees in Inner London* (1988) showed that affordable housing in both the rented and freehold sectors was fast becoming a rare commodity, particularly for residents of Central London.

Recent changes in the planning 'use classes' have meant that access to a balanced range of employment opportunities is also becoming difficult to achieve. Already threatened by land market pressures, premises and jobs are now subject to increased pressures as developers and owners seek changes to higher-value uses to an even greater extent than previously. Not surprisingly, this has added to the multiplicity of problems. The index of deprivation based on the 1981 Census spotlights Central London as having many of the wards with the highest scores. The net result of these pressures and side effects has been to weaken the community structure of Central London, a structure which has long provided an identity on which to build.

Faced with this situation, community organizations in Central London in 1980 formed themselves into a new organization, the Campaign for Homes in Central London (CHiCL). CHiCL is an umbrella organization comprising locally based groups who are in turn umbrella organizations for the Central London communities of Fitzrovia, Paddington, Spitalfields, Bermondsey, Hammersmith, Waterloo, Covent Garden, Battersea, Finsbury, Shoreditch, Somers Town, King's Cross, Pimlico and Soho. All the member groups share a common problem – their location on the fringe of the central commercial core of London (CHiCL 1986).

Whilst each organization was initially involved in its own local campaign, having become aware of the need to share experiences and pool information it proved only a short step for these groups under the CHiCL umbrella to start to develop policy initiatives to tackle some of the problems facing them collectively. From putting forward alternative plans at public inquiries into large development schemes, this evolved into proposing modification to existing plans, at both the local and London level.

One of the first initiatives by CHiCL was the 'LifeBelt' policy. As a result of the changes and pressures talked about earlier, the idea of creating a zone to protect communities around the central commercial core from unbalanced development was proposed. Based on the green belt concept, the LifeBelt was innovative for two important reasons; it drew attention for the first time to the problems being experienced by Central London communities, and it demonstrated that the problem was being felt by a large number of communities over a wide area of Central London, and that only a strategic response was likely to be adequate.

The ideas behind the LifeBelt were taken up by the Greater London Council (GLC) during its 1981–6 administration. What emerged was both a policy response, and substantial resources to back it up in the form of what the GLC called its 'Community Areas Policy' (CAP). The CAP did not exactly mirror the LifeBelt concept, opting instead for a slightly wider definition of areas of need than that of Central London communities (see Figure 9.2). Nevertheless, the main focus of the CAP was a ring of designated communities around the central commercial core, based on those working under the CHiCL umbrella. The CAP was developed over

Figure 9.2 GLC's 'Community Areas'

the period of the GLC administration, and subsequently by a mixture of community-based organizations and London Borough Councils. Something in excess of £30 million was allocated to the Policy by the GLC planning committee and other committees of the Council (Greater London Council 1985).

The CAP was responsible for funding more than 150 projects, perhaps the best known among them being the Coin Street and Courage's Bottling Plant sites in Waterloo and Bermondsey, and the Battlebridge Basin site in King's Cross. It is important to stress that the CAP was not only cash-based. Of equal significance was the development of new planning policies for London. This was through the long overdue process of updating the GLDP, which was completed in 1984 (Greater London Council 1984). The revised plan included the concept of Community Areas, along with many other new planning policy initiatives.

In the event, as mentioned earlier, the amended GLDP became a casualty of the abolition of the GLC. Although the strategic support for the CAP was withdrawn, CHiCL was instrumental in ensuring that a number of boroughs embraced the concept in their local plans. Since abolition, CHiCL and its membership have been striving to maintain an influence on the emerging planning policies being drawn up for London by LPAC, the government and the boroughs. This has met with some success. The 'strategic advice' submitted by LPAC to the DoE in 1988 during the process of the government preparing its 'strategic guidance' (DoE 1989) contained as one of the components of its 'four-fold vision of London', 'a city of stable and secure residential neighbourhoods capable of sustained community development'. This advice also contained the concept of 'Community Areas of Need' a direct derivative of the GLC's Community Areas Policy. These are areas mostly in and around Central London which LPAC feels 'should be given a priority to secure them as residential communities and neighbourhoods from commercial pressures, and provide accommodation for services to meet community development needs'. It is left up to the boroughs to identify such areas. The government's guidance also highlights the importance of the community idea. In its introduction the government stresses the importance of ensuring that 'local community life can flourish' and points out that 'One of London's particular strengths is the distinctive identity and character of the many localities and communities which together constitute London. Unitary Development Plans (UDPs) should reflect this local diversity and vitality'. Quite how this will be interpreted by the boroughs remains to be seen, but both the government and LPAC seem to have indicated their acceptance of the concepts being pursued by CHiCL and its member groups.

Overall then the ideas which first emanated from community groups have not only provided the stimulus for strategic policy development over a prolonged period but have also led to a number of key projects which still provide the valuable model for community development first expounded by Abercrombie – that the planning task is largely a matter of 'defining, completing and reclaiming communities'.

THE NEIGHBOURHOOD IDEA

Throughout this century there has been an ongoing argument between those who advocate that the neighbourhood is the essential organ of an integrated city, and others who brand the whole idea as the creation of romantic sociologists. Opponents of the neighbourhood idea often portray it as being in conflict with the need to look at the city as a whole. This suggests a model whereby the component parts of a city draw their identity and role from what is perceived to be in the interests of the city in its widest sense. In practice this means the parts being sacrificed to the good of the whole –a classic top-down model of decision-making. Supporters of the neighbourhood idea do not accept this approach. Neither do they accept that such a conflict exists. It is their view that the city is fundamentally the sum of its parts, and it is this which gives it its traditions, vitality, identity and strength. It is on this vision of the city that supporters of the neighbourhood idea pin their hopes for the survival of the city. It is those who see the city as a mere machine who champion the first approach. It is a lover of the city who champions the second. The GLC's Community Areas Policy was undoubtedly a move towards re-introducing the neighbourhood idea into planning practice. What community organizations at the time didn't know was that this was not the first time that the master plan for the Capital had stressed the importance of the community structure of London.

Undoubtedly, the high point of neighbourhood planning was in the period immediately following the Second World War. Sir Patrick Abercrombie wrote in the Greater London Plan that 'the social arguments for community planning are now fairly generally accepted'. People like Clarence Perry, Raymond Unwin, Charles Cooley and others vigorously promoted the concept. A counter-movement also existed, mainly in the shape of Reginald Isaac, but others like Frederick Osborne were known not to be enthusiastic about the idea.

Whatever these differences in relation to neighbourhood planning, they were all heavily influenced by even earlier advocates of a new order in the planning of cities. People such as Robert Owen and James Silk Buckingham proposed a number of model solutions in an effort to confront what they saw as the evils propagated by the industrial revolution. They, and others, were heavily influenced by the moral improvement movement of the day. Others saw in the city the chance to create an ideal form. Both of these strands of thinking could be categorized as being essentially top-down in nature. Certainly advocates of the garden city and new town concepts were firmly in this camp. In that sense what Abercrombie and others proposed were more 'planned communities' than 'community planning', something which was to emerge much later.

It is worth reminding ourselves of the ideas that underpin neighbourhood or community planning, because they are shared by advocates of both, even if the practice differs. Lewis Mumford was obviously very sure of his ground when he stated that 'the neighbourhood is a social fact' (1954). This was based partly on his own work, as well as that of others who had studied the neighbourhood structure of cities in some depth. Raymond Unwin was equally emphatic about the importance of 'fostering the feeling of local unity in an area'. He thought it important 'to see how far it was possible for the growing city to secure an end so desirable as the greater localization of life' (1920–1). Abercrombie picked up on this theme, when he proposed that 'each community would have a life and character of its own, yet its individuality would be in harmony with the complex form, life and activities of the city as a whole' (Abercrombie 1945).

The concept of 'knowability' has also been developed by Hillier and Hanson (1984), who put forward the proposition that 'by embodying intelligibility in special form, individuals in society create an experimental reality through which they can retrieve a description of their society and the ways they are members of it'. They are very critical of the way planners and architects have often proposed schemes that have led to 'social fragmentation'. Another key person in the history of developing a 'theory of good city form' was Kevin Lynch, who in his seminal work (1960) put forward the idea of 'imageability', that is the capacity to evoke a strong sense of place in the observer. His idea was that a highly image-able city would be one that, 'could be apprehended over time as a pattern of high continuity, with many distinctive parts clearly connected'. Lynch also stressed the importance of structure and continuity.

This connection between the neighbourhood and the city in its totality was something which was certainly understood in the medieval period. Thinking then saw the city as a whole, being expressed through the city's cathedral spire. Through what is now called 'design dispersal', this expression was then echoed throughout the city by the neighbourhoods and their civic and religious build-ings. The neighbourhood idea, or any idea which gives order to a city introduces a similar organizing force, which over time leads to a solidification of the design idea. In contrast to these principles of buildings and cities deriving from land design, architects like Le Corbusier prided themselves on 'the great amputation', a system of design which specifically set out to divorce building design from the discipline of land design. Enormous damage has been done to cities as a result of this concept of 'painting on a clean canvas'.

Yet another idea which underpins the idea of neighbourhood or community is its rôle in preserving individuality. This is essentially a defence against the trend towards what has been called 'mass society'. But equally, as Jacqueline Scherer pointed out (1972) the concept allows 'structured commonality', or to put it another way 'the ability to create an identity from inside'.

PLANNED COMMUNITIES OR COMMUNITY PLANNING?

The idea of neighbourhood or community has therefore been considered over a long period. Discussion in the past has been very much framed within a philo-sophy which sees cities simply as sophisticated machines, or products of a design idea. Undoubtedly a city based on past concepts of neighbourhood would be preferable to one based on many other city and anti-city theories that have found favour over the last decades. Whilst it is a vital first step to explore concepts of city structure and form, it is important to take the discussion further and build a 'theory of city process' into the equation. Essentially, the neighbourhood as developed by Abercrombie and others was top-down in its conception. This is why it is important to make a distinction between 'planned communities' and 'community planning'. Planned communities such as those proposed by Abercrombie, bold and innovative as they were at the time, were nevertheless a product of the school of ideas which tended to see the city as a machine, albeit a human one. It was a philosophy which held that the application of a common set of standards would produce a desired end product. The problem with this approach is that it is inherently rigid. It encourages the idea of a model solution, and as we all know from bitter experience, model solutions produce model problems. It is an unfortunate fact that once new ideas have been adopted and established planners and architects often proceed to apply them rigidly, whether to a green-field site or to a city centre, to an old or a new town, to a western or eastern city. It is this rigidity of application that has made it imperative to constantly find new ways of refreshing the planning and architectural professions. A good example of this rigidity was pointed out by Mumford (1954) when he accused planners of 'sacrificing sociability and concentration to mere openness'. More recently, Hillier and Hanson (1984) returned to this theme, accusing urban designers of 'de-spacializing society and de-socializing space'. The nature of cities, particularly old cities, requires something more organic and humanistic if the subtle balance between past and future is to be achieved.

This is where community planning comes into play. Community planning does not start from a set of standards, or have as its ultimate objective an idealized goal. It starts from a belief in planning as an on-going *process*, one which is people-focused and is based on a set of *principles* rather than standards. Key amongst those principles are the twin concepts of 'identity' and 'stability'. These twin pillars of community planning are what the ideas discussed in this chapter effectively boil down to. They provide the bed-rock on which all other elements of community planning are built. Community Planning then is aimed at reinforcing rather than undermining the idea of a 'common memory', encouraging rather than diluting the intensification of civic energy, promoting rather than frustrating communication and ensuring rather than obstructing continuity.

CONSIDERING A NEW APPROACH TO COMMUNITY PLANNING

Having looked at the idea of neighbourhood and community from the standpoints of early planners and also described certain policy initiatives taken by community organizations in London, it is time to turn to how community planning as practised is evolving, and whether it is possible to construct a theory of community planning. In other words is there a set of principles and practices which are sufficiently distinct and robust to warrant a place in the lexicon of planning practice? It may not yet be possible to answer that question but there are grounds for thinking it might emerge.

It has already been suggested that previous notions of neighbourhood were 'top-down' in conception. In a sense the Community Areas Policy (CAP) was from the same stable, although it was formulated as a direct result of citizens' action. When community organizations in London first started out on the road of taking a direct interest in planning, it was primarily a defensive reaction. Only later did ideas of becoming involved in city-wide planning start to develop. In the nature of how things develop organically, it was the ideas of the green belt and those contained within the Greater London Development Plan (GLDP) that became the focus of attention for community groups, rather than those of Abercrombie and earlier pioneers of the concept of neighbourhood. This was not because of any deliberate decision on their part, merely because those ideas were still to be uncovered.

Essentially the CAP was a derivative partly of the rather vague zoning concept by which the GLDP designated areas of London as 'preferred location' for office or industrial development, and partly of the principle inherent in the green-belt concept. Of course it was no accident that areas in Inner London designated as preferred office locations were those where community groups first became active. It was the very fact of the designation as a preferred location that in turn led to continued development pressure. Having successfully made the case for action to the Greater London Council (GLC), as strategic planning authority, the GLC were then faced with translating the commitment into policy. The simplest route, and in the event the one the GLC took, was to amend an existing document. Hence it was only a short step for the areas designated in the GLDP as 'preferred office locations', to become 'preferred community locations' or 'Community Areas' as they then became known. But this was more a matter of convenience than design and a case of fitting theory to structures, rather than the other way round. Creating a policy umbrella which sought to protect existing communities from development pressures requires in the long run something more sophisticated than a mere zoning mechanism. This is because it only represents one part of the planning equation. Having the protection of such an umbrella is crucial to enabling the growth and development of community-led initiatives. But for lasting stability, what is required is a change in the way planning and urban design is perceived and practised at every level, whether developers, local authorities, architects or community organizations. It is the belief of those promoting community planning that such a shift can only occur if planning or plan-making has as its ultimate objective the identification and support of the community of a city. However, for this to be sustained it requires some kind of more fine-grained mechanism than a broad zoning tool like the CAP, or the other rather mechanistic concepts of neighbourhood developed by Abercrombie and others. It is also necessary to recognize that while some communities will generate a tremendous variety of different activity, others

129

might be slower to react, yet more respond too late, or even in some cases not at all. Equally, it is possible that ideas on community once conceived and built into a plan in a structured way, might lead to a uniformity of approach which risks doing harm to the residents of a city. This seems to suggest a need to fuse the ideas of community planning and planned communities.

Where do we start? One of the features of the British system of local government has been its lack of stability over the last 30 years. This has been caused by the number of local government reorganizations. Quite apart from this lack of stability, the process of change has left British people substantially worse represented than other European residents. If one takes the French commune as a bench mark, it can be seen that the ratio of local representation to residents is 69 times worse in the UK than in France. The UK is drastically out of line with other European countries too. The process of amalgamations inherent in past reorganizations has had the effect of loosening local ties, in the process undermining the significance of the historic county, borough, ward or parish structure. This weakens in turn citizens' ability to influence decisions, and undermines the identity ties which people have developed over generations. The 1964 and 1974 reorganizations were of course followed in 1986 by the abolition of the Greater London Council and the metropolitan counties, shortly to be joined by the abolition of the historic county structure of England and Wales if the proposals from the Labour and Conservative Parties are to be believed. The result? The twin pillars of the neighbourhood idea 'stability' and 'identity' have become casualties of the British fetish of searching for the perfect structure.

In the process more and more power has been placed in the hands of politicians and professionals. In contrast the principles behind community planning are that people should have control over their own destiny, be involved in decisions that affect their lives, and be treated with respect by those who make decisions on behalf of other people. This is not just a lofty principle; it is founded on the belief that people themselves represent an enormous resource simply waiting to be released and tapped. It is also based on an understanding of the proper rôle of politicians and professional people. Politicians need to understand better that their rôle is as 'intermediaries'. Similarly professional people need to see themselves as 'enablers'. At present there is a tendency for both to see themselves as 'controllers'. This is a crucial distinction to think about and understand, if the hidden resource that people represent is to stand any chance of being unlocked. A shift in approach will only flow from an understanding of the distinction between people as citizens and people as subjects.

Another feature of the last 30 years has been a move away from plans drawn up in the style of Abercrombie towards plans based on little more than statements of intent. Plans have therefore become much less spatial in their content and conception. Whereas the idea of community was evident in the Abercrombie plans, later plans have become little more than development control documents. Present plans in the UK are almost totally devoid of any 'vision'.

Community planning is then a reaction against these long-standing trends.

Recently, in an effort to start to put some flesh on the bones of a new approach to community planning, CHiCL and the North Southwark Community Development Group have embarked on a new project called *Your Place Comes First* (Maclean-Thorn and Nicholson 1991). The aim of the project is twofold: first to reinstate the community structure into plan-making and decision-making, second to develop a new tool to equip those involved in the process of preparing development proposals and plan-making with an ability to analyse their proposals more comprehensively. In particular the project will investigate the links between locality and community and examine the nature of community consciousness and the way in which citizens can define their own communities. In short it is hoped the project will give a little polish to the Abercrombie idea of 'defining, completing and re-claiming communities'.

The question may be asked as to what kind of electoral arrangement would be necessary to sustain community planning. I have always been critical of the tendency in the past to concentrate on structure rather than process. In a sense once the community structure is accepted as the basic building block, then the rest of the infrastructure of decision-making should fall into place. However, there are grounds for deep concern over the current fashion for unitary authorities.

I have no doubt that the local government reorganizations of the past were wrong-headed and deeply damaging to the twin ideas of people being in control of those decisions that affect their lives, and the concept of community. The damage will in part be repaired if the current wave of enthusiasm towards decentralization is sustained. The unitary authority on the other hand posed a real threat to ideas of community and community planning because it is an inherently rigid system. Simplistic notions of devolving services to the lowest tier of government demonstrate an extraordinary naïvety about the dynamics of power. Ask almost anyone who has been involved in community campaigns and they will tell you that the problem usually starts with the failure of the lowest tier adequately to recognize either the community organization or the issue facing them. This was certainly the case in Central London. The lesson community organizations have learnt through bitter experience is that it is absolutely crucial to have structures which allow *alternative* solutions and *alternative* escape routes for those facing problems. Unitary authorities threaten to prevent both. Add to this the fact that one of the main criticisms of plans and planning in the past has been the lack of progress with implementation and one has to conclude that what is required is *more* people involved in implementing projects, not fewer.

It is essential that ordinary people are recognized as the hidden resource (Labour Party 1991) they undoubtedly are. From here it is a logical next step to see them as a key focus for the implementation of plans and projects. Community planning offers a vehicle for unlocking that resource. 'Model' organizations can then be developed to give expression to this vast pool of energy, vision and ideas. The Coin Street project (Nicholson 1988) on London's South Bank provides one such model which takes the form of a non-profit company of community builders (see Figure 9.3). The concept of community trusts provides

131

Figure 9.3 Coin Street development – family housing and open space in the heart of commercial London

another. There are undoubtedly others which exist and still more that will be developed in response to a particular need or project. One thing is certain. It is that people will ultimately prevail even if it has to be despite rather than because of the structure of government that happens to be in fashion.

What this chapter has tried to do is spell out the way community organizations in London have developed the idea of community planning. It has explored earlier attempts to introduce the concept of community planning into the infrastructure of planning. It has also set out the bones of a new research project designed to take a further step down the path towards developing a theory of community planning. It is my firm belief that it is only when the principles behind community planning are incorporated into planning practice that it will become a process through which there is some hope of giving expression to Castells' dream that 'at last citizens will make cities' (Castells 1983).

REFERENCES

Abercrombie, P. (1945) *Greater London Plan: 1944*, London, HMSO.

Bacon, E. N. (1967) *Design of Cities*, London, Thames & Hudson.

Campaign for Homes in Central London (1986) *City Life – a Future for Central London*, London, CHiCL.

Campaign for Homes in Central London, with Brendan Hodges (1988) *Access to Housing: the Problems for Employers and Potential Employees*, London, Bartlett School of Architecture and Planning, University College London.

Castells, M. (1983) *The City and the Grassroots*, London, Edward Arnold.

Docklands Consultative Committee (1990) *The Dockland Experiment: A Critical Review of Eight years of the LDDC*, London, DCC.

DoE (1989) *Strategic Planning Guidance for London*, London, HMSO.

Greater London Council (1984) *Alterations to the Greater London Development Plan*, London, GLC.

—— (1985) *Community Areas Policy – a Record of Achievement*, London, GLC.

Hillier, B. and Hanson, J. (1984) *The Social Logic of Space*, Cambridge, Cambridge University Press.

Labour Party (1991) *New Opportunities for the Inner Cities – Labour's Strategy for Urban Renewal*, London, Labour Party.

Lynch, K. (1960) *The Image of the City*, Cambridge, MA, Massachusetts Institute of Technology Press.

Maclean-Thorn, D. and Nicholson, G. (1991) *Your Place Comes First*, London, North Southwark Community Group.

Mumford, L. (1954) 'The neighbourhood and the neighbourhood unit', *Town Planning Review* 24 (4): 256–70.

Nicholson, G. (1988) 'City as commodity or community', *Architects' Journal* 187, 30 March, pp. 34–52.

Scherer, J. (1972) *Contemporary Community*, London, Tavistock Publications.

Unwin, R. (1920–1) 'Distribution', *Town Planning Institute: Papers and Discussions* 7: 37–45.

10

GOVERNING THE CAPITAL

Michael Hebbert

The main thing wrong with London today is its structure of government. The late Greater London Council had few friends, but its spiteful, impulsive abolition in 1986 – a Nietzschean folly Brendan O'Leary (1987) called it – left a system that has fewer. Agreement on this point is so widespread that we may be in danger both of exaggerating the defects of the current system and expecting too much of proposed alternatives such as the Labour Party's Greater London Authority (Labour Party 1991).

Just what is wrong with London government today? Back in 1985 many thought, and some even hoped (Sofer 1987:64), that the decapitation of metropolitan government would cause urban chaos. There was a hint of disappointment in County Hall when a GLC-funded forecasting project found that the most likely scenario was not a breakdown of services but a gentler process of 'policy drift' in which standards slipped and new problems were left unsolved (Greater London Group 1985). Six years later it might seem that the pessimists were right and the LSE forecasters wrong. The message of this book is not slippage but *crisis*.

We are in one of those bouts of self-disgust that periodically afflict all major cities. A sense of teetering on the brink runs through this book and through much recent journalism – the *Evening Standard Magazine* ('The London debate', 15 December 1989), the *Architectural Journal* ('A Vision for London', 14 March 1990), *Christian Action Journal* (another 'Vision for London', Autumn 1990), the *Sunday Times* ('Open letter on the capital crisis', 30 September 1990), LWT's *London Programme* ('London into Europe: the challenge ahead', 9 November 1990) and the same station's *London Lecture* (Sir Ralf Dahrendorf, 'Does London need to be governed?', 6 December 1990). The unifying theme is one of civic decline. London's sordid trains and the homeless in their cardboard boxes become symbols of the breakdown of government in a metropolis bereft of proper institutions. Andrew Neil of the *Sunday Times*, his grammar collapsing under the strain, expressed this despairing sense of an institutional vacuum in an editorial letter to the Secretary of State:

No-one governs London ... what's to be done? Roads are seizing up. For those living in the city crime, vandalism and litter are at their worst ... In quiet residential streets there seems to be virtually no control over what people can do to ruin their houses and the appearance of whole streets ... There is a strong feeling that there is nobody in control, that London, once one of the world's best-planned and best-run cities, has run out of control.

Anxiety about civic decline has an international dimension. London's aimless squalor is repeatedly contrasted with the brisk efficiency with which Paris builds its infrastructure, cleans its streets, promotes its image and woos the international investor (e.g. Henley Centre 1990). The argument for reform is underscored by warnings that London will 'lose the race' with Frankfurt and Paris to become financial and business capital in the single European market. 'Whilst other European cities make bold and adventurous plans for the 21st century, London is threatened with catastrophe simply because no-one in government cares enough about the capital's future' (Greater London Labour Party 1991:4).

It is worth thinking twice about the London/Paris comparison, which has become rather clichéd. Mayor Chirac of Paris certainly has more political clout than a London borough leader, and a clearer strategic vision of his city. The 4,000 Parisian street-cleaners in their bright green boilersuits are impressive. But we must be careful to compare like with like. Both the administration and the geography of Paris have always been centralized. It is the archetypal concentrated city, shaped by the authority of state prefects. London has a different history, in which local-level councils pursued 'their rights of self-government almost up to the point of urban anarchy' (Lees 1973:426). Paris grew by unified civic design, London by higgledy-piggledy expansion around many centres, sprawling outwards into an assemblage that as Paul Theroux puts it (cited by Elliott 1986:2), more resembles an entire country than a single city. Paris has always been proudly conscious of its destiny as the cultural capital of Europe, if not the 'point of equilibrium' for the entire world (Evenson 1984:258). For London, on the other hand, European leadership is a novel obsession: apart from an imperialist phase before the First World War, London has not nursed claims to be capital of anything other than England, or at most the United Kingdom.

London's greatest chronicler shared the view that things were better ordered across the Channel. 'The shabbiness of our English capital as compared with Paris', wrote Charles Dickens in 1868,

> I find very striking after an absence of any duration in foreign parts. The meanness of Regent Street set against the great line of Boulevarts [sic] in Paris is as striking as the abortive ugliness of Trafalgar Square set against the gallant beauty of the Place de la Concorde. No Englishman knows what gaslight is until he sees the Rue de Rivoli and the Palais Royal after dark.
>
> (1868:454)

Of course, he meant central Paris. There were few gas lights in the outskirts, or sewers either. Then as now there was a tendency for visitors to Paris, even such inquisitive visitors as Dickens, to be beguiled by its brilliance and panache and miss the underside of poverty in the *banlieu*. London's shabbiness has always been more public, its virtues more homely and private, tucked into an irregular patchwork of back gardens, squares, parks, small terraced streets, and family homes (see Rasmussen 1982: Olsen 1986).

Critics tend to be concerned less with the history of Parisian consistency and London variety than with the obvious present contrasts in the way the two cities are run: Paris has clean streets, Disneyland and a mayor (so the argument runs), London has dirty streets and no Disneyland, *ergo* London needs a mayor. This logic may be faulty. The local government arrangements set up by the 1985 Act have some bearing on London's problems. But just how much?

Take 'cardboard city'. It sprang up after GLC abolition yet has little to do with the transfer of former GLC housing responsibilities to the boroughs. It is a direct product of government policy at the national level – the withdrawal of benefits from 16–17-year-olds living away from home, the narrowing of entitlement to social security benefits under the 1986 Social Security Act, the restriction on the Social Fund in 1989, the freezing of child benefit, and the decade-long immobilization of the public housing sector. Other factors have played their part – especially the failure of 'community care' to cater for all former patients released by the closure of the big psychiatric and subnormality hospitals. Like other elements of the crisis described by fellow-contributors, sleeping rough is not specific to the metropolis but reflects mainstream government policy during the Thatcher years. The dislocations figure more prominently in London because of the city's size and its disproportionate shares of the richest and poorest segments of the populations, whom the politics of Thatcherite redistribution most favoured/penalized, London being a gigantic lens on British society as a whole.

National rather than local factors explain the ignominious comparison between strategic planning in London and Paris. The planning of any capital city is a paper exercise unless it has the involvement and backing of central government which is – directly or indirectly – the paymaster for national infrastructures that, by definition, converge upon the capital. Bold planning in Paris is not a product of local government arrangements: it reached its apogee under Napoleon III when Paris had no local government at all, and to this day is driven as much by presidential and prime ministerial as by mayoral authority. In contrast the administrative fiascos of the Channel Tunnel terminal, and transport in Docklands, stem directly from refusals by central government to fulfil the responsibility it assumed under the 1985 Local Government Act for the strategic planning of London. It is not just the local government framework that is wrong – the example of Paris points precisely towards the need for centrally led inter-ministerial initiative – so much as the government's adamant ideological conviction that major land-use and transport decisions are best left to the vagaries of the market.

Any discussion of the shortcomings of transport planning tends to lead back to the 'race for leadership' between London, Frankfurt and Paris. There is no straightforward connection between London's ranking as a decision centre, and the fragmentation of its local government arrangements. As far as financial services are concerned, the most important factor is the regulatory and fiscal environment. London's success in capturing European business in the late 1980s was a direct result of 'Big Bang' deregulation – nothing to do with local government. Its lead was bound to be eroded as Frankfurt, Paris, Milan and other financial centres updated their trading systems.

A factor of secondary importance is the office stock. This local government can influence, for better or worse. There is an element of truth in the argument that London's competitiveness on this front was enhanced by the abolition of the metropolitan authority in 1986. The reasons lie in the survival of the City Corporation within the square mile of the central financial district. Throughout the past century, the existence of the City as a separate jurisdiction has had a palpable effect of distancing London-wide policy-makers, first in the London County Council and then in the Greater London Council, from the work of the financial and business services sectors. Their requirements for modern office space were consistently downgraded by strategic planners, who did what they could to dampen new office development or disperse it out of Central London. The GLC's proposed amendments to the Greater London Development Plan of 1984, and its London Industrial Strategy of 1985, gave overriding priority to the manufacturing sector and had little to say about office development – except that existing jobs should be protected against automation, and new office development should either be confined to the existing sites or located in secondary centres such as Barking, Brixton and Walthamstow (GLC 1984:28–32; 1985:333–59). The consequence of a negative policy, sustained over many years, was that London office space was cramped and expensive, and badly designed too, since developers were confident of a seller's market. London's potential to win business as a global financial centre, based on history, a favourable system of regulation and the convenience of the English language, was chiefly constrained by physical floorspace.

All that changed rapidly in the later 1980s. Freed of the braking effect of the upper tier, London ran through an office boom which has substantially enlarged and upgraded the building stock and shifted the market from chronic under-supply into surplus. True, infrastructure development has lagged behind property development, especially in the Docklands, but that lag will be made up by projects now in the pipeline such as the Jubilee Line extension and the cross-rail link. London's lack of a single metropolitan authority is an incidental factor for the financial markets; its office stock, on the other hand, matters a good deal.

The more we probe London's problems, the less easy it is to pin them down to the 1985 Local Government Act. GLC abolition happened to coincide with a period when London performed strongly on the tangible variables such as GDP, job-creation, business start-ups, property values, construction activity and migration. Londoners in regular employment could reasonably take a stoic attitude to the by-products of economic success – road-works, noise pollution, travel congestion – that diminished the quality of their everyday life. The electorate was given the opportunity to express its view on the structural question in the local elections of May 1990. The Labour Party campaigned on a 'crisis of London' ticket:

> Ours is the only major capital in the world with no city-wide government of its own ... Now we are in danger of being left behind as the great cities of Europe adapt to a changing world. London urgently needs new ideas and positive policies to stop the slide, otherwise hard on the heels of a declining quality of life will come a decline in business and industry. There are ominous signs that this is happening already.
>
> (Labour Party 1990)

A fresh start in governing London headed the list of the party's electoral promises in the May election. Voters were unresponsive. While Labour enjoyed a 5 per cent swing nationally (since the Euroelection of June 1989) there was a 4 per cent swing *against* the party in London. Local variations, for instance against Labour in Brent and for it in Lewisham, indicated that voters cared about the quality of local government at the borough level, but the issue of city-wide provision rang no bell.

There is evidence that the economic downturn has brought a shift in attitudes and a growing sense amongst Londoners – outer as well as inner – that a new structure is needed for the metropolitan area as a whole. The favourable press and poll coverage when the Labour Party unveiled its proposals for a Greater London Authority in the summer of 1991 confirmed suspicions amongst parliamentarians on both sides of the House that balkanization is perceived as an issue in London and there are votes to be won in tackling it. The issue is not – as I now want to show – that people are dissatisfied with the 33 boroughs, but that something is needed over and above borough government. London as a whole, within the green belt, is a tangible community. People belong to it as well as to Lambeth or Barnet. Unitary, one-tier government, still a fad of the post-Thatcher Conservative Party (DoE 1991), may perhaps be appropriate in free-standing towns but is an unsuitable model of local democracy for big cities, and particularly London, where there is no eponymous core municipality (like the cities of Glasgow or Birmingham) that can claim to speak for the wider whole. Large-scale and local levels equally need expression. Reinstating a metropolitan government will not be a panacea for the problems of the homeless or the commuter. Tackling issues on the ground will always depend as much on the government of the day, and on external factors, as on the interventions of a metropolitan authority. But with no representative body to articulate them, all problems hang loose in a political limbo. Just as a nation-state needs its institutional shell to watch and safeguard common territorial interests, so do big cities. There is, after all, a real link between balkanization and civic decline. Voters feel it in London today.

Since further reorganization is unavoidable, what form should it take? The Labour Party's proposals are colourfully argued but contain little concrete analysis (Greater London Labour Party 1991; LP 1991). The Conservative position, at least officially, is to bury its head in the sand: 'the Government has no plans to change the general structure of local government in London' (DoE 1991:7). It may be helpful to review the present arrangements, looking at strengths as well as weaknesses.

The process of reattributing all the functions of the GLC and ILEA took several years. Legislation laid down the framework of a London government system based on 33 unitary authorities, but much of the detail was left to be resolved by negotiation between the boroughs, central government, and the GLC's 'ghost' incarnated in the London Residuary Body. The powers, personnel and assets of the top tier were split three ways (Hebbert and Travers 1988). Some functions, for example, strategic planning and traffic management on main roads, went upwards to enlarged London divisions within the Departments of the Environment and Transport. A small number have gone to quangos (e.g. English Heritage, London Arts Board, South Bank Board, National Rivers Authority) or to the private sector (e.g. the former computing and scientific services of the GLC). But the greater part remains under local political control through joint committees of boroughs. The major shared services, of which the largest is the London Fire and Civil Defence Authority with an annual budget of £200 millions, are only the tip of the very large iceberg of borough collaboration (see Table 10.1). Many of these joint services are administered through 'lead boroughs' on behalf of London in general, involving officers as well as elected members in substantial external commitments.

It's worth considering the case for joint action. Some political scientists of the new right, looking at the performance of metropolitan governments in American cities, have argued the virtues of running big cities through a lot of small units instead of one big one. Drawing parallels with economic markets, they argue that co-ordination can as well be achieved by decentralization as by centralization on a large-scale authority (Bish 1971; Ostrom et al. 1961). Mrs Thatcher was influenced by new right thinking in many policy fields, but not apparently in the abolition of the GLC and metropolitan counties, which had no clear rationale at the time and has never been justified by Conservative intellectuals in terms of a pure public choice strategy (O'Leary 1987; Thornley 1991). Yet there can hardly be a better example of a polycentric metropolitan system than the 33 separate local authorities who make up London government. We are not dealing here with tightly-bounded, exclusionary, parasitical suburbs that are the principal beneficiaries of metropolitan fragmentation in the United States. Leaving aside the City Corporation, the London boroughs are large, evenly matched units, with populations in the range of 130,000–300,000. The majority are focused on substantial shopping and service centres and all combine housing and employment and span at least some of the immense social and physical variety of the capital city. The borough councils are powerful general-purpose

139

Table 10.1 Joint bodies with London Borough membership

Concessionary Fares Scheme
Coroners' Courts Committees
Docklands Consultative Committee
Docklands Transport Steering Group
Greater London Enterprise
Home Loans Portfolio
Inner London Education Advisory Committee
Inner London Special Education Forum
London Ambulance Service
London Boroughs Children's Regional Planning Committee
London Boroughs Disability Committee
London Boroughs Emergency Planning Committee
London Boroughs Grants Committee
London Boroughs Training Committee
London Canals Committee
London Committee on Accessible Transport
London Ecology Committee
London Fire and Civil Defence Authority
London Planning Advisory Committee
London Helicopter Advisory Committee
London Road Safety Committee
London Recycling Forum
London Research Centre
London River Association
London Tourist Board
London Waste Regulation Authority
Magistrates' Courts
SERPLAN
Waste Disposal Authorities (7)

authorities with current budgets of £100 to £300 millions. Ever since they were created under the 1963 London Government Act, the centre of gravity has been their town halls, not County Hall (Self 1971). In complete contrast to Paris, there is no dominant core authority. The Cities of London and Westminster, and seven adjacent boroughs, divide the core between them in a pattern that reflects London's historic variety and many-centredness. Despite its initial unpopularity the division into 33 parts has become ingrained by force of habit and the accumulated effect of borough decision-making (Hebbert 1991). The Local Government Boundary Commission (LGBC) has been reviewing London since 1987. Any real debate over the division of the capital was bound to surface in the course of this exercise. Despite widespread press publicity all that the Commissioners received on the general pattern of London authorities were two insignificant submissions from individual members of the public (LGBC 1991:3). No sign of crisis here.

140

A polycentric system of metropolitan government is only feasible if there is co-operation between the units. Before 1965 London local government had been riven by intense rivalries and local hostilities. Attempts to solve common problems by joint action broke down repeatedly, chronic failure of voluntary co-operation counting for more than any other factor in the decision to create the GLC (Herbert 1960; Smallwood 1965; Rhodes 1970; Davies 1988). The story has been different since 1986. Party-political difference rather than localism is now what most divides the 33 councils. New left and new right have both used London boroughs as testbeds, the polarization reaching its climax at the time of GLC abolition when the Labour-controlled councils broke away from the London Boroughs Association to form their own Association of London Author-ities. But here is the surprising thing. Despite the division of the polycentric system into two more or less ideologically opposed blocs the boroughs have managed to maintain a wide array of services on a co-operative basis. Maybe, using a geopolitical analogy, the blocs have helped to channel conflicts and stabilize the system as a whole. Individual boroughs have found it easier to co-ordinate since acquiring unitary status, because they can 'deliver' without intervention by a higher authority. Their size and equality has assisted joint working. The traditional free-riders in the metropolitan system, Conservative-controlled suburban councils (Young and Garside, 1982), have taken a much more active interest in capital affairs since becoming unitary London Boroughs instead of reluctant victims of annexation within a two-tier 'Greater London'.

If there is a 'crisis' in London government today it is not one of internecine strife at the borough level. The main litigation involving boroughs has been over residual GLC and ILEA matters such as Coin Street and Hendon Aerodrome. Joint boards have found themselves able to agree budgets and develop policy, with the notable exception of the London Boroughs Grants Committee, which was saddled with an unworkable voting system. The various reserve powers taken by the Secretary of State in the 1985 Act to intervene and resolve inter-borough conflict have been little used. As predicted by Ostrom et al. (1961:616), the mere threat of intervention by a higher-tier authority may have sufficed for the members of a polycentric system to agree to subordinate their differences in order to maintain local control. Some of the Secretary of State's most contro-versial recent interventions in London, particularly in land use planning, have been made against a common front of all boroughs. While overt political co-operation between LBA and ALA is the exception rather than the rule, the three main issues so far pursued – on homelessness, access to the European Social Fund and parking enforcement by local council wardens – have all yielded tangible results. The associations have also worked together over the transfer of education from ILEA to Inner London boroughs. Government ministers are growing accustomed to joint delegations of chairmen of the respective LBA and ALA committees to press London's needs. At a routine level London-wide linkages of technical officers are a well-established and important feature of the system. Chief officer associations exist for all the main functions performed by

the 33 councils. Most meet monthly and provide a forum for exchanging information and articulating shared London-wide interests through discreet channels of
technical influence. Informally, they provide professional networks that can
handle cross-boundary issues. The political parties also provide an important
infrastructure, completely lacking in New York, for cross-London policy issues.

This is not a rosy view of London government. The boroughs include some of
the best and some of the worst local authorities in Britain and the variation in
service quality has been stretched by regressive national policies of redistribution
between the two population segments most overrepresented in London, the very
poor and very rich. But taking London as a whole, the boroughs and the City
have proved a workable and stable polycentric system. The past five years have
seen a tangible growth of civic pride in the borough centres that is curiously at
odds with this book's message of a capital in crisis. Outer London centres such as
Kingston and Ilford have been pedestrianized and planted to provide urban
environments that compare with attractive provincial towns. In Inner London
there has been a shift from the esoteric and clientelistic concerns of the 'new left'
towards street-level service improvements: cleaning, caretaking, maintenance.
This emphasis on environmental quality is as evident in Labour-controlled
Lewisham and Islington as in Liberal Democrat Tower Hamlets or Conservative
Wandsworth.

Wheeled dustbins in Lewisham, van-based litter patrols in Wandsworth, and
the transformed street scene of Tower Hamlets, where utilitarian concrete street-
lights have been replaced by elegant dark blue lamp standards in cast iron, all
bear witness to another feature of a polycentric system predicted by Bish and
Ostrom – its stimulus to innovation (see also Jones and Stewart 1983). One of
the first things to happen after GLC abolition was a rash of new boundary posts
with borough names and slogans: 'Westminster – the heart of London',
'Wandsworth – the brighter borough'. The signs themselves are unnecessary
street clutter, but they express a competitive spirit which has encouraged
boroughs to be experimental and learn from each other. A few examples of innovation that have proved influential are Redbridge's appointment of a Town
Manager for central Ilford, Islington's reorganization of client services into
district 'one-stop shops', Havering's promotion of kerbside collections for
recycling, Sutton's environmental audit, Brent's use of private leasing to house
the homeless, Westminster's parking wardens and the initiatives in light rail
investment being made by Croydon and Kingston.

Today's crisis, then, is not in the structure of borough government but in its
superstructure. The system loads too much onto joint arrangements, leaving local
politicians overstretched by the accumulation of political duties in the ward, at
borough level, and for London as a whole. There is an unhealthy tendency for
councillors to deputize for each other on joint committees, whose shifting
membership weakens democratic control. The traditional criticism of decentralized systems is that they are prey to localism. It has not been too much in

evidence in London, but there is perhaps a tendency to overemphasize geographical parity in resource allocation, whether for investment in the fire brigade or grants to the voluntary sector. The system penalizes investments that cross boundaries, such as bus-lanes, and favours ones that are implementable internally, such as town-centre pedestrianization. Co-operative decision-making breaks down when faced by challenges such as the great vacant swathes of railway and industrial land straddling borough boundaries in the middle ring of London – Stratford, Park Royal, Old Oak Common. Above all, borough politicians remain accountable to borough electorates. It is not their job to take the wider view of London's interests. It is nobody's job.

The 1985 Act's decentralization to the joint or individual boroughs was accompanied by a less visible, and more damaging centralization at the national level. Central government has greatly increased its power of administrative intervention in the capital, but not its sectoral structures of decision-making. At least 1,000 civil servants are now working full time on London topics inside the Departments of Transport, Environment, Education, Employment, Health and the Home Office, but within narrowly departmental terms of reference and in physical isolation from each other. The bizarre separation of land use and transport elements within the *Strategic Planning Guidance for London* (DoE 1989) illustrates the problem well. It might matter less if there were political will in Parliament and the Cabinet to care for London, but both are weak. London's 85 MPs find no common ground. The capital is perhaps the most disadvantaged region in Britain in terms of lobbying power within the House of Commons. The Cabinet is dominated by the departmental interests of its members, augmented in the past decade by the distinctive aversion of Thatcherism to any cross-sectoral planning. Occasional short-term exceptions have been made to the government's 'hands-off' rule. One is the Docklands Transport Steering Group set up by Mrs Thatcher, reputedly in response to strong personal lobbying by the Reichman brothers, developers of Canary Wharf. It brings together central departments, the utilities, the police, the boroughs and major private developers and is just the sort of strategic policy group needed to cut the Gordian knot of the Channel Tunnel terminal and the major rail investments elsewhere in London. However, government has been at pains to set no precedent.

The debate about running the capital continues, and seems likely to come to a head as soon as there is a change of government or a major compound disaster that exposes the precariousness of current arrangements whereby the Department of the Environment London Division provides emergency co-ordination between boroughs, police forces (City, Metropolitan and Transport), British Telecom, London Fire and Civil Defence Authority, and the London Ambulance Service. This chapter shows why the focus of concern has been strategic, not local. The challenge of any future reform is to build on the positive basis of borough government with a metropolitan superstructure that encroaches as little as possible on their autonomy.

Various structures are on the drawing-board. Tony Banks MP and Sir Ralph

Dahrendorf (1991) have argued that it should take the form of a metropolitan executive headed by an elected mayor – a new idea in British local government – who would draw down powers currently in the hands of central government and quangos. Lady Porter (1990) proposes that it should be internalized within government through a Ministry for London, under legislative supervision of a Grand Committee of London MPs. Andrew Coulson for the Fabians (1990), the Greater London Labour Party (1991) and the Labour Party (1991) want to insert a directly elected strategic authority for London as part of a general constitutional scheme for devolution to regional and national assemblies. The Labour Party incidentally sees the creation of a London Metropolitan Council as the opportunity to carry out its long-standing commitment to abolish the City Corporation and to divide it among the adjacent boroughs (Greater London Labour Party 1991:38; Labour Party 1991:5).

Other suggestions have a more incremental character, involving changes within existing legislative and institutional frameworks. Quasi-federal characteristics already present in the system could be strengthened by giving joint bodies stronger executive powers. The London Fire and Civil Defence Authority could, for example, be given civil emergency planning powers, resuming the lead rôle in disaster prevention and management formerly performed by the GLC. The London Planning Advisory Committee could become a planning authority, and take on the research and intelligence role currently with a separate joint committee, the London Research Centre. In the longer run an accumulation of functions in the hands of joint committees might lead to the re-establishment of a top tier of general-purpose government, but indirectly rather than directly elected so as to maintain the unitary principle. The 'metropolitan communities' in Montreal and (formerly) Toronto offer practical precedents of city-wide agencies with responsibility for policing, physical planning and infrastructure, accountable to their member municipalities rather than directly to the metropolitan electorate.

Sceptics will reply that a strengthened borough federation could never survive for a long run. Voluntary co-operation occurs today because the joint bodies are relatively weak. The base would crumble if it were built upon. Opponents of re-centralization in the London government system would simply bide their time until a further change of government enabled reform to be undone again.

It is worth considering one last strategy for the capital which might accommodate both sceptics and reformers. It focuses on the Sleeping Beauty of the London government system, the City Corporation. The one common feature of all reforms since the mid-nineteenth century is that they have left the City untouched. A unique medieval relic, it survives with its Norman boundaries, its Lord Mayor at Mansion House, its Sheriffs, Aldermen and Common Council in the Guildhall, its police force, and its small electorate of freeholders intact. Though this chapter refers to '33' London boroughs, the polycentric system really has 32-plus-1 elements, and that one appears to have more in common with the City Dickens knew than with its modern partners.

The past few years have brought important changes for the City. It has lost its historical monopoly over banking and financial services. The magic of the 'square mile' evaporated in the 1980s as major office developments spilled for the first time across the City boundary into Tower Hamlets, Hackney, Islington, Camden, Westminster and across the river to Lambeth and Southwark. It seems likely that the immense scale of Canary Wharf will, in the long run, downgrade the City into one of several nodes in a linear central business district. The City Corporation was able to survive all previous rounds of local government reorganization because of the peculiar function of the square mile within the national economy. It was the goose that laid the golden eggs – but that uniqueness has now gone.

At the same time, the City's distinctive electoral structure has been exposed to new scrutiny by a historical accident – the expansion and merger of City-based business services, especially accountancy, law and financial consultancy, largely as a by-product of privatization. Firms in the business service sector are based on the model of professional partnerships rather than the limited liability company. *Partners* in firms with freehold premises in the square mile qualify for the electoral roll where *directors* do not, and they can vote in all wards where their firms have premises. The electoral roll (the 'Ward List') increased from 15,000 to 18,000 electors in 1990 because of multiple registration induced by mergers. Elections for Common Council are now dominated by firms such as Coopers & Lybrand Deloitte, which has 700 partners, and premises in several of the electoral wards. Much larger companies, however, remain disenfranchised. The City's electoral arrangements have always been anomalous, but recent trends are that much harder to defend.

The City's rôle as a local government has also been changing in recent years. Its wealth, centrality and studied political neutrality made it an obvious home for stray London-wide functions after GLC and ILEA abolition. In a short period the City Corporation acquired responsibility for managing and maintaining over 2,500 traffic lights in the capital, together with their computerized control systems; the ownership of Hampstead Heath; the Greater London Record Office, London's central historical archive for London; the public health service at Heathrow Airport and the Port of London; and the careers service for Inner London schools. With County Hall no longer available for the reception of visitors, Mansion House has played a greater part in providing civic hospitality on the capital's behalf. The Corporation has found itself in an unaccustomed leadership rôle in lobbying for infrastructure investment. Through these and other means the City's historical isolationism has been broken down, willy-nilly, by GLC and ILEA abolition. Though only one of 33, it has become *primus inter pares* in a polycentric London government system.

The renaissance of the City Corporation would be shortlived if Labour were to carry out its proposal to abolish it and divide the square mile amongst the adjacent boroughs. However, the party has clearly not thought this one through. Abolition offers little benefit at a high political cost. The division of the spoils

145

would be immensely controversial. It would be seen (as Mrs Thatcher's abolition of the Greater London Council is seen) as a wanton and vindictive act, and the City's ghost would haunt any successor arrangement.

There are better ways forward. In 1935 Herbert Morrison concluded his book *How Greater London is Governed* with a chapter called 'The Might-Have-Been'. It was a brief, wistful rumination on the opportunities lost when the City Corporation failed to enlarge its boundaries and functions in step with London's eighteenth- and nineteenth-century expansion. Instead it clung tenaciously to its ancient municipal constitution and privileges, leaving the management of modern London to be taken in hand by Morrison's own empire, the London County Council based on County Hall, and predecessor authorities such as the London Schools Board and Metropolitan Board of Works. The cradle of British freedom and local democracy (which Morrison rightly saw the City to be) became the most prominent anomaly in the local government system:

> we can only sigh our regrets that the great City of London Corporation did not develop and expand as it might naturally have been expected to; that it has become largely shut off from the pulsating life of the far-flung London of the twentieth century; and that Greater London does not share in the enjoyment of the rich historical traditions, the civic glory and the dignity of the first British Municipal Corporation, and of Guildhall and the Mansion House.
>
> (Morrison 1935:160)

Today Morrison's whimsical might-have-been takes on a new interest. His beloved County Hall is empty and up for sale like any other surplus office block. The London County Council has given way to the Greater London Council and its abolition in turn has left an institutional vacuum which neither the modern boroughs nor central government and its quangos adequately fill. Meanwhile the City Corporation has suffered some erosion of its traditional basis, but increased its standing as a 'lead borough' within a wider polycentric system of government in the capital. There is a basis for radical evolution here. The City's governing body – the Common Council – could be modified to represent London as a whole, either by direct election, perhaps on the basis of Euro-constituencies, or better, by indirect election through the borough councils. Its functional rôle could be enlarged to take responsibility for traffic and transport as well as traffic lights, planning and economic development as well as the wholesale markets, current research as well as historical records, external promotion as well as the Lord Mayor's Show. A redistribution of functions to London's oldest and proudest local government might prove less partisan and so more durable than any new creation. It would restore a political identity to London while leaving that of the boroughs intact.

Perhaps we are straying too far into the realm of What-Might-Be. For the time being we have a polycentric system of borough government. Though further reform is inevitable, the gist of this chapter is that the system has more positive qualities than *Sunday Times* readers have been led to believe. The challenge is to design a superstructure that respects those qualities. Morrison closed his book on a note of uncertainty that will do as well for today:

> Whether the future will witness a straightening out of the tangle of London government consistent with the principles of democracy I do not know.
> I hope so.
>
> (1935:173)

ACKNOWLEDGEMENT

The author thanks John Hall for kind advice and the three loyal colleagues who read the chapter in draft – June Burnham, Tony Travers and George Jones – for comments and corrections. He bears sole responsibility for any remaining errors of fact or interpretation.

REFERENCES

Bish, R. (1971)*The Public Economy of Metropolitan Areas*, Chicago, Markham.
Cheshire, P. (1990) 'The outlook for development in London', *Land Development Studies* 7:41–54.
Coulson, A. (1990) *Devolving Power: The Case for Regional Government*, London, Fabian Society.
Dahrendorf, Sir R. (1991) *Does London Need to be Governed?* (script of LWT Lecture broadcast 6 December 1990), London, London Weekend Television.
Davies, J. (1988) *Reforming London: The London Government Problem 1855–1900*, Oxford, Clarendon Press.
Dickens, C. (1868) 'The boiled beef of Old England' from *The Uncommercial Traveller* in *Collected Works of Charles Dickens* vol. XIII, London, Gresham Publishing.
DoE (1989) *Strategic Planning Guidance for London*, London, Department of the Environment.
—— (1991) *Local Government Review: the Structure of Local Government in England – a Consultation Paper*, London, Department of the Environment.
Dunleavy, P. and O'Leary, B. (1987) *Theories of the State*, London, Macmillan.
Elliott, M. (1986) *Heartbeat London: The Anatomy of a Supercity*, London, Firethorn Press.
Evenson, N. (1984) 'Paris 1890–1940', in A. Sutcliffe (ed.) *Metropolis 1890–1940*, London, Mansell.
GLC (1984) *The Greater London Development Plan – as proposed to be altered by the Greater London Council*, London, Greater London Council.
—— (1985) *The London Industrial Strategy*, London, Greater London Council.
Greater London Group (1985) *The Future of London Government*, London, London School of Economics.
Greater London Labour Party (1991) *Report of the Commission of Enquiry into the Future of London's Government*, London, Greater London Labour Party.
Hebbert, M. (1991) 'The borough effect in London's geography', in D. Green and K.

Hoggart, *London: A New Metropolitan Geography*, London, Edward Arnold, 191–206.

Hebbert, M. and Travers, T. (1988) *The London Government Handbook*, London, Cassell.

Henley Centre (1990) *London 2000: A Report Prepared for the ALA*, London, Association of London Authorities.

Herbert, E. (1960) *Royal Commission on Local Government in Greater London 1957–60 – Report* (chairman Sir Edwin Herbert), Cmnd 1164, London, HMSO.

Jones, G. W. and Stewart, J. (1983) *The Case for Local Government*, London, George Allen & Unwin.

Labour Party (1990) *London Pride*, London, The Labour Party.

—— (1991) *London – a World Class Capital*, London, The Labour Party.

Lees, L. (1973) 'Metropolitan types – London and Paris compared', in H.J. Dyos and M. Wolff (eds) *The Victorian City: Images and Realities*, London, Routledge & Kegan Paul, 259–87.

LGBC (1991) *Report No. 594: Review of Greater London, the London Boroughs and the City of London – the London Borough of Barnet*, London, Local Government Boundary Commission for England.

Morrison, H. (1935) *How Greater London is Governed*, London, People's Universities Press.

O'Leary, B. (1987) 'Why was the GLC abolished?', *International Journal of Urban & Regional Research* 11(2):214.

Olsen, D. (1986) *The City as a Work of Art*, Yale, Yale University Press.

Ostrom, V., Tiebout, C. M. and Warren, R. (1961) 'The organization of government in metropolitan areas – a theoretical inquiry', *American Political Science Review* 55(4):607–30.

Porter, Lady S. (1990) *A Minister for London: A Capital Concept*, London, FPL Financial Limited.

Rasmussen, S. E. (1982) *London – the Unique City: Revised Edition*, Cambridge, Mass., Massachusetts Institute of Technology Press.

Rhodes, G. (1970) *The Government of London: The Struggle for Reform*, London, Weidenfield & Nicholson.

Savitch, H. (1988) *Post Industrial Cities: Politics and Planning in New York, Paris and London*, Princeton, NJ, Princeton University Press.

Self, P. J. O. (1971) *Metropolitan Planning* (Greater London Paper no. 14), London, London School of Economics.

Smallwood, F. (1965) *Greater London: the Politics of Metropolitan Reform*, Indianapolis, Bobbs-Merrill.

Sofer, A. (1987) *The London Left Takeover*, London, J. Caslake.

Thornley, A. (1991) *Urban Planning under Thatcherism*, London, Routledge.

Young, K. and Garside, P. (1982) *Metropolitan London: Politics and Urban Change 1837–1981*, London, Edward Arnold.

11

DOCKLANDS: DREAM OR DISASTER?

Andy Coupland

There are a number of major developments in London which make the headlines: the development of King's Cross, the Channel Tunnel terminal saga, the redevelopment around St Paul's Cathedral. None however have received the scale of coverage – or represent the national significance – of the development of London's docklands. Developments in the area have been controversial for over twenty years, but the past decade has seen an accelerated pace of change and a dramatic new direction for an area traditionally viewed as one of the poorest and least attractive for investment.

Docklands has been, for over a decade, an experiment. The new Conservative government in 1979 came to power with two fundamentally different approaches to developing the Docklands area from that taken over the previous five or six years. First, public sector involvement was inappropriate; what was needed was a market-led approach. Following from this was a belief that the improvement of the area would come from the 'trickle down' of benefits from this market-led redevelopment. Planning had a very limited rôle to play; the market was to determine what should be built and where it should be built. The public sector's involvement (headed by a Board of private businessmen appointed by the Secretary of State) was to 'prime the pump' of development by taking land from reluctant public landowners and preparing it for private sector development. This involved spending money to build infrastructure and transport facilities and to prepare land for construction. It also involved marketing and advertising the benefits of the area to attract development interest.

The LDDC (the London Docklands Development Corporation, set up to 'regenerate' Docklands) has come under fierce criticism at times for underestimating the possible success which it might achieve, as much as for the nature of the development which has been built. However, much of the criticism should at least be shared by the government who set it up. The Enterprise Zone established by the government in part of Docklands in the Isle of Dogs can be viewed as a machine, given a series of complex instructions and then having its controls sealed up for ten years. Those instructions say, effectively, that the area will have planning permission, that developers who spend money can subtract it from their company or personal tax bills and that the government will pay occupying companies' rates for the ten years that the Enterprise Zone will be in existence.

Once switched on, the machine cannot be switched off, and the controls cannot be altered. The LDDC set about creating a new transport infrastructure which they thought would be appropriate to open the area up and meet the needs of the potential space which might be created. Working jointly with the GLC the LDDC assessed the possible benefits of constructing the Docklands Light Railway. In 1982 they concluded that there could be up to 2 million square feet of office space developed as a result of opening the DLR, and progressed the scheme accordingly at a cost of some £77m. for just over 7 miles of track. By the time the railway opened in 1986 the 1,500-passenger-per-hour capacity was already being stretched. The scale of development had increased so much that nearly 20 million square feet of offices were expected to be built in the Isle of Dogs alone – a figure which has now reached some 26 million square feet. As the DLR became obsolete before it opened, so other transport schemes, particularly roads, had to be provided. From a few local distributer roads costing a few million pounds the LDDC found the need to create an entire new East London transport network.

The original idea was that the LDDC would 'prime the pump', and then gradually need less and less public money as the private sector took over. In practice, as more and more private capital was committed the LDDC had to find more and more money to create basic infrastructure, and was then criticized for not having provided it soon enough. The criticisms were not just from outside; former LDDC Board member Wyndham Thomas described the LDDC's approach as characterized by a 'woeful disregard for commonsense planning practice, such as ... keeping infrastructure provision (especially public transport) in line with development' (quoted in Docklands Consultative Commitee 1990).

From a budget of £50m. to £60m. in the early years of the Corporation they received over £300m. in 1990–1, and many further millions were committed to Docklands in transport schemes, and through the Treasury paying Business Rates or through lost tax revenue due to Enterprise Zone development. And despite all this development activity, local unemployment is still higher than when the LDDC was created, and few benefits can be seen to have 'trickled down' to the local communities.

BEFORE THE LDDC

London's docks were built from the early nineteenth century, and developed steadily through to the 1930s. They were created in areas outside the City, often on marshy riverside peninsulas like the Isle of Dogs, although sometimes their construction involved disruption of the existing community, as at St Katharine Docks where 1,250 houses were cleared away, leaving 11,000 homeless. Once built, and surrounded by walls up to 20 feet high, the docks became private places, separate from the communities of seamen and dockworkers who lived around them. The communities were poor and often living in overcrowded and unpleasant conditions.

By the mid-1960s a combination of underinvestment, a change to larger ships unable to use limited dock capacity and a realization of the potential for the redevelopment of the area led to the closure of the first upstream docks. This process continued through the 1970s until the last docks closed – the Royal Group of docks in 1981.

The first of the docks to close, the St Katharine Docks, were sold to developer Taylor Woodrow in 1969. A prime site opposite the Tower of London held obvious development potential, and with a hotel and office development the scheme was soon under way. However, it was far less obvious what should be done with the other docks should they become available for redevelopment. In 1971 the Conservative Secretary of State for the Environment, Peter Walker, appointed consultants Travers Morgan to carry out a comprehensive study of Docklands.

Rather than create a single blueprint the consultants were asked to come up with a range of options. The main features of these were: *Waterside*, a linked water park with new housing, 75 per cent for sale, new north–south roads, and a Holiday Hotel for the Royal Docks. *Thames Park* linked large areas of open space, with some new housing split equally between sale and renting. *East End Consolidated* was a token attempt to show continuing industrial and public rented development in the area. *City New Town* was, as its title suggests, a major development, split again equally between rented and private housing with a major entertainment and shopping centre and significant office development in a parkland setting. Finally there was *Europa*, perhaps the most interesting in hind sight because of its proposed major injection of private housing, and a rapid transit system to link office and shopping centres.

As a result of the consultation process related to these proposals, and in order to try to protect the remaining jobs in the docks, a number of local Docklands Action Groups were formed throughout the area to campaign for a different approach to the development of Docklands. Local Labour councils were very unhappy with a scheme proposed by a Tory government and GLC, and following a win by Labour in the GLC elections in 1973 the Secretary of State, Geoffrey Rippon, withdrew the proposals and tried a new approach.

Instead a joint committee, with the Borough Councils and the GLC as well as the Port of London Authority and TUC delegates was proposed. This was to co-ordinate development rather than usurp the powers of the local councils. This Docklands Joint Committee (DJC) and its officers (including from 1978 the promotional Docklands Development Organization) set out to consult the public, partly through a Docklands Forum of community representatives who were given first one and later a second seat on the DJC, and through the publi-cation of a series of planning documents culminating in the London Docklands Strategic Plan of 1976.

While the policies contained within this plan were 'evolutionary rather than

151

revolutionary', the overall objective was clearly different from that which had guided the Travers Morgan studies.

The main objective of the London Docklands Strategic Plan was:

> To use the opportunity provided by the large areas of London's dockland becoming available for development to redress the housing, social, environmental, employment/economic and communications deficiencies of the docklands area and the parent boroughs and thereby to provide the freedom for similar improvements throughout East and Inner London.
>
> (Docklands Joint Committee 1976)

The plan contained a whole range of proposals, including housing density and tenure guidelines, shopping centres, transport improvements and community facilities. The DJC set to work on implementing the strategy, but a number of problems soon became apparent. Land assembly was difficult, especially as many public authorities like the Gas Board, Port of London Authority or British Rail were reluctant to give up their holdings. The supply of financial support slowed, especially for transport developments reliant on central government money – which IMF controls on spending were making increasingly difficult to obtain.

Progress was made, but only slowly. Land was prepared, several hundred houses were built, and a whole range of community facilities were programmed and commenced. In addition private development, especially industrial schemes, were coming forward. Among these were the London Industrial Park in Newham, the move of Billingsgate Fish Market and the creation of the Cannon Workshops in the Isle of Dogs, and the construction of the News International printing plant in Wapping. The St Katharine Docks continued to develop, particularly as a centre of office development.

The DJC barely had 3 years to implement the plan before the 1979 election brought a Conservative government with a very different, and as was soon to be seen, a very committed view of how Docklands should be developed.

THE ARRIVAL OF THE LDDC

The new Secretary of State for the Environment, Michael Heseltine, wasted no time in announcing a new organization, with sweeping powers, to develop Docklands. What was new about such an organization was that, rather than carrying out any development of its own, it was to facilitate the private sector's activity in the area.

The LDDC also took a very different approach to planning – from the start the approach has been 'flexible' – in the words of the LDDC's Chief Executive, Reg Ward early in the life of the Corporation, 'I don't believe in the planning system. I'm opposed to land-use planning, quality does not flow from it' (1981) and later he said, 'we are making a virtue of having no grand strategy' (1982). These are not isolated quotations; Ward was appointed by Michael Heseltine to the LDDC job precisely because of his views.

Similarly, Nigel Broackes (now Sir Nigel Broackes, then as now of developers Trafalgar House), the first Chairman of the LDDC, expressed a hostility to a traditional planned approach. 'I do not intend that we should have a rigid plan to which developers must conform' (1981). This 'market-led' approach can be found in all the LDDC's activity, and many of its statements – published and unpublished – to date.

The LDDC was quickly seen to be a very different organization from the DJC. Indeed, when it was set up it was unique in the amount of power given to a body accountable only to Parliament, and not to any local electorate. The LDDC is the planning authority for development control in Docklands (although it has no powers to make statutory local plans). Initially it received around £60m. every year from the government. The Secretary of State was able to 'vest' land in the LDDC previously owned by any local authority or nationalized body. At first around 1,000 acres of land were acquired by or vested in the Corporation. This represented most of the land holdings of all three borough councils, plus large areas formerly owned by British Gas and the Port of London Authority (PLA). Later a further 1,000 or so acres were acquired so that virtually all the developable land in the 5,000-acre Docklands area passed through LDDC ownership, including the Royal Docks in Newham, which were bought by the LDDC on a long lease early in 1986.

The Board of the LDDC is appointed by the Secretary of State, meets in private (with no press or public present) and publishes no agendas, minutes or decisions (although the planning committee now meet in public). The LDDC was described by the Secretary of State for the Environment in 1980 as a 'single-minded' development agency 'with the sole task of regeneration' (Heseltine 1980). Just what the LDDC means by 'regeneration' has become clearer over the past few years.

It involves a very different set of objectives from those spelled out in the London Docklands Strategic Plan. Although the LDDC was required to take the LDSP as its 'starting point', as Nigel Broackes admitted to a House of Commons Select Committee, the LDDC never intended to take the contents of the plan seriously as a guide to their activities.

The powers available to the corporation allow it to 'carry out any business or undertaking for the purpose of regeneration', although no clear definition of regeneration has been published. This gives the LDDC almost limitless powers. In addition the LDDC is the planning authority for Docklands. Planning applications are submitted to the Corporation, and local councils have a short period (usually 14 days) to make comments, which the LDDC may (and frequently does) choose to ignore.

The LDDC controls all aspects of development control within the Docklands area, but plan-making powers reside with the local authority. This was in order that development in Docklands should be broadly in accord with the views of the local authorities. However, the only statutory plan in existence at the start of the LDDC's control of the area was the Beckton Plan. This was breached a number of times, with the Secretary of State refusing to intervene. The Tower Hamlets Plan has since been adopted, and is being ignored by the LDDC in a number of respects. In Southwark the Secretary of State refused to allow the council to adopt their local plan without changing it to accommodate the views of the LDDC.

In general, relationships between the councils and the LDDC have been very poor. The LDDC's attitude to the local community has consistently been to ignore them if it seems likely that their view might in any way block progress of significant development – however valid the community concern might be. As a marketing agency the LDDC is selling development land, and would therefore prefer to present the image of 'wide open spaces' rather than an area with over 40,000 inhabitants, with their own views on how development should proceed.

Early in its history the LDDC decided to try to buy off those parts of the community which it could get over to its side. A report to the LDDC Board revealed this approach: 'Money will help. By doling out the equivalent of urban aid in an efficient, friendly and generous way, the LDDC will win some support. More or less, people do appreciate the hand that feeds them' (LDDC 1982a).

The LDDC has sometimes published documents which they refer to as 'area strategies', but which they admitted (in private at least) were 'in all but name local plans' (1982b). However, as was also admitted, these documents have never been changed in any respect as a result of the 'consultation exercise' which has accompanied the publication of some of them. At no time in the first four years after it was set up did the LDDC hold a public meeting, anywhere in Docklands, to explain any of its plans for the area. Even local councils have found it impossible to discover exactly what is going on, and one major development proposal was exhibited in Switzerland before it was even announced in this country!

THE ENTERPRISE ZONE

The Isle of Dogs Enterprise Zone covers 170 hectares (425 acres, or less than 10 per cent of Docklands total area) in the boroughs of Tower Hamlets and Newham. Most of the land is around the Millwall and West India Docks on the Isle of Dogs. Ironically, just as Geoffrey Howe announced the initiative for Enterprise Zones in his budget speech of March 1980, a consensus had been reached between local residents and Tower Hamlets Council as to the future of the area. The 'Isle of Dogs Local Plan' filled out the broad provisions of the London Docklands Strategic Plan. The northern section of the Island was to be developed for industry, the south predominantly for housing – mostly council houses with gardens so that two isolated council estates of predominantly high-rise blocks could be linked together. Industry already on the Island was to be encouraged to stay and expand.

When the boundary of the Enterprise Zone was drawn it was designed to exclude all existing riverside industry (most of it now approved for housing by the LDDC) and developments new to the area such as Billingsgate market and the Asda superstore. On the vacant sites in the area, the LDDC approved (as landowners – no planning permission being necessary) a series of developments. At first these were predominantly industrial, but they have rapidly increased in scale and size so that by 1986 around 4 million square feet of offices were being developed. Many of the original smaller industrial buildings were converted to office use. Others were redeveloped, sometimes only a matter of months after their completion, to make way for larger schemes, invariably of office development. By 1991 there were plans for over 25 million square feet of office space; nearly a third of the City of London. This is in complete opposition to the intentions of the original plan.

The creation of Enterprise Zones was based on the idea that businesses were being stifled by controls and needed liberating. It was the fashionable view that 'red tape' was killing initiative. The aim was therefore to create a 'free-fire' zone within which capital could operate unhindered.

Within an Enterprise Zone:

(a) Developers can offset their building costs against Income Tax or Corporation Tax. The taxpayer thereby subsidizes the construction costs of any buildings.

(b) There are no planning controls. An authority is designated under the legislation as the 'zone authority', but normal planning permission is not required. This allows flexible use of buildings. Shells can be constructed and the use modified to suit the most profitable market. From the developer's point of view it therefore helps minimize risk and maximize profit.

(c) Businesses are exempt from rates for the 10 years' duration of the Enterprise Zone. The government pays the rates direct to the local authority. While this may not directly profit the developer it does mean that higher than average rents can be charged, which increases the value of the development. Rates were described by Patrick Jenkin (briefly Secretary of State for the Environment in the mid-1980s) as 'the one tax businesses can't avoid', which may also explain the attraction of this concession.

Visitors in the early days of the Enterprise Zone could see new developments starting to take shape within only a few years. What was more difficult to see was the low level of activity that takes place in those buildings. Advanced Textile Products had an impressively designed modern warehouse, but it was a fully computerized operation, employing only one person. Since then the walls have been removed and windows substituted (with no planning permission being necessary) and the building is available as an office.

Initially the marketing strategy of the LDDC seemed based on duplicating the sunrise belt of the M4 in Docklands – a suburban office park with wind-surfing at lunchtime. Hi-tech offices and services were the image. Manufacturing industry was not encouraged and often displaced: it is not what the tourists or bankers want to see. Rents quickly became too high for most manufacturing firms. Only two developments in the Enterprise Zone have been priced low enough to attract the kind of industrial development local people could expect to compete for jobs in: Cannon Workshops and Riverpark. Both are on land which was retained by the Port of London Authority and hence not attributable to the LDDC. Cannon Workshops was partially demolished for the construction of the massive Canary Wharf office development.

The basis of the LDDC's approach was to prepare land for private developers. It cleared derelict buildings, improved services and utilities and installed new infrastructure. The sites were often disposed of at a price well below the value of the publicly financed works. While land for housing was obtaining prices of well over £2m. an acre by the end of the 1980s, sites for office development in the Enterprise Zone continued to be offered at less than £1m.

CANARY WHARF

This project has featured in the media above all others in Docklands – and rightly so (Figure 11.1). The largest development undertaken by a single developer anywhere in Europe, it consists (in its contemporary version) of nearly 12 million square feet of commercial space. The scheme has had a chequered history. Originally it was expected to cost some £1.5 billion, and was proposed by a consortium of US banks, fronted by a developer called G. Ware Travelstead. His master plan received the backing of the LDDC in 1985, displacing a number of more modest proposals for the central quay and surrounding docks in the centre of the West India Docks. However, by 1987, despite attracting support from Crédit Suisse, First Boston and Morgan Stanley International, the scheme was running into problems. Detailed designs were worked up, and land acquired, but the necessary finance had not been identified. At this stage the entire scheme was taken over by Olympia & York, a Toronto-based development company, which already owned some 45 million square feet of office space in the US and Canada – including 24 million square feet in New York.

Plans were revised, additional architects brought in and commitments confirmed to provide transport infrastructure. In May 1988 the first piles were sunk, and the largest building in phase 1, the 800-feet tall tower, was topped out in November 1990. As the project is located in the Isle of Dogs Enterprise Zone it normally requires no planning permission. However, the taller buildings are more than 50 metres tall, and so a form of permission is required and part of the site, at the western end, adjacent to the River Thames, is not in the Enterprise Zone. The main negotiations with the LDDC took place in relation to their ownership of the major part of the site. This led to the creation of a Master Building Agreement, which includes clauses on the employment of local labour, as well as the provision of transport improvements.

The project has been controversial from its inception. It introduces the tallest building in the UK into a sensitive location, particularly when viewed from Greenwich. It creates office space for an estimated 40,000–60,000 workers in an

Figure 11.1 Canary Wharf. Courtesy of Olympia & York Canary Wharf Ltd

area with limited transport and road infrastructure. It offers clerical and managerial jobs in an area with high levels of unemployment, particularly among the unskilled and semi-skilled workforce. It represents a potentially massive financial contribution from public funds through the operation of capital allowances in the Enterprise Zone. It shifts the centre of development eastwards, creating a wholly new business district equivalent in size to a city centre, 2 miles from the City of London. Current costs are estimated at over £3 billion.

Initial negotiations with the Travelstead consortium identified the need to extend the Docklands Light Railway into the City, at Bank, to link to the existing underground railway network. This project was taken on by Olympia & York, who have committed £88m. to the cost of the new link and improvements to the original system. This will allow a tenfold increase in capacity of the railway – only completed in 1987 – by 1993. The LDDC and Department of Transport have initiated a series of road projects to improve access to the Isle of Dogs, including the £300m.+ Limehouse Link tunnel to the west and the Lower Lea

Crossing to the east. In the summer of 1990 London Underground announced that the Jubilee Line would be extended from Green Park through Canary Wharf to Stratford. Olympia & York will contribute some £400 million of the cost of this scheme.

Despite delays in construction, the main part of Canary Wharf's first phase was completed in 1991. Landscaping commenced in December 1990, when the first mature trees (over 30 ft tall) were installed around the Westferry Circus roundabout (a two-storey structure). Various options for later phases of development and associated projects in the Enterprise Zone are still being considered by the developers. American architects Ehrenkrantz and Eckstut have drawn up plans, as have Koetter, Kim Associates of Boston. Sir Norman Foster may design one of the later-phase towers. However, the developers have indicated that they have decided not to proceed with later phases until the lettings market shows a significant improvement.

Olympia & York, working with development partners Trafalgar House, started work on Port East, an adjacent site which includes a range of Grade 1 listed warehouses which will be developed for retailing, restaurants and bars, while adjoining land will have offices and a hotel. Plans are being drawn up by Olympia & York with Imperial Land and the Tianjin government of China for a long-promised scheme at Poplar Dock, also in the Enterprise Zone, where a 7-acre site is to be developed. The developers also own over half the adjacent Heron Quays site, where the adjacent owners of the remaining part of the site, Tarmac Brookglade, have produced plans for further high-rise office development designed by Scott Brownrig & Turner.

HOUSING IN DOCKLANDS

Covering a wider area than the LDDC, the DJC's proposals were to build 23,000 new houses to add to the 19,000 already in the area. The tenure mix was to be 20 per cent owner occupation, 30 to 40 per cent for various forms of equity shared ownership, and 40 to 50 per cent for rent. These figures represented the ability of the local population to acquire property, and the need to build houses with gardens to make up for the huge numbers of flats constructed in the 1960s and 1970s. The London Docklands Strategic Plan commented that owner occupation was 'beyond the financial grasp of all but a minority of those who do not own a dwelling'.

While some progress was made in the few years the organization had to develop the schemes, the main efforts were put into preparing land, which was quickly taken over by the LDDC when it took control of Docklands.

The Conservatives' concerns, translated by the LDDC into action, were with the tenure of the housing in the Docklands boroughs. Less than 10 per cent of housing was owner-occupied; the government's view was that the area was therefore 'unbalanced'. The LDDC was given a clear indication that it was expected to support private housing developers, and not intervene in questions of housing need. Its main rôle in relation to the existing population has been to ensure nomination rights for local tenants on schemes built on LDDC land. This means that local residents have a short period – a few weeks – to put their name down for new housing schemes.

The degree of government control and influence over construction of housing in Docklands became increasingly apparent throughout the 1980s. The most striking example has been seen in the question of finance to allow local authorities to build housing for rent. The LDDC has offered to make land available to the councils provided they could guarantee to start construction within a certain period – usually one year. This proved all but impossible for all the local authorities as their Housing Investment Programmes have been progressively cut by the government. As a result almost no new local authority housing has been built since the LDDC was established. In addition, through the vesting process, most of the local authority land reserves were also taken by the LDDC.

During the House of Lords hearings which established the LDDC in 1981, the Chairman elect, Nigel Broackes, gave evidence on the LDDC's proposed housing programme. He outlined the intention to build 8,000–10,000 dwellings up to 1991. Of these 50 per cent would be for sale, 25 per cent to rent and 25 per cent available on an 'equity shared' basis. Only a year later, however, the LDDC identified 75–85 per cent of housing on the land it owned as being for sale, and by October of 1982 the LDDC was stating in its (unpublished) 'Corporate plan' that '10 per cent of new homes could be available on a rented basis, i.e. about 200 units a year'.

In fact, in the first five years of the LDDC considerably fewer than 1,000 units of rented accommodation were provided, and there was a net loss of rented accommodation as councils, faced with unreasonable repair bills for older blocks, are selling them off either for demolition or for refurbishment by the private sector. Even where local councils did consider the possibility of building houses to rent they found it increasingly difficult to acquire land as land values rose.

One of the LDDC's objectives was to raise the value of the land, and this has been achieved as values have risen from £56,000 an acre in Beckton to 'ten times that' (*Chartered Surveyor Weekly* 17 May 1984). Prices climbed to over a million pounds an acre for riverside sites, and Southwark Council was asked to pay £1m. an acre for a site for which it was compensated at £80,000 an acre when it was vested in the Corporation four years earlier.

It is the LDDC's job not only to sell the idea of Docklands, but to sell Docklands itself. Owning over 20 per cent of the land this has proved lucrative, with the Corporation paying on average £50,000 an acre. Initial schemes were sold as 'loss leaders', as explained above, but by the mid-1980s land prices had climbed dramatically, and were rising into £1m. or more per acre. Unlike councils, the LDDC can keep and spend all the money it makes, so this is worth several million pounds a year in additional income. The developer is getting a good deal too, as the LDDC may spend many times the final price in preparing the land. For example, a site in the Isle of Dogs, London Yard, was purchased by the LDDC for

159

£222,000. £2,256,000 was spent on a river-wall, decontamination and site development of the 7-acre site, and it was then sold for £818,400. The developer then built 296 houses and flats at prices which started at £120,000 (and rose steadily over the next few years).

The LDDC argued initially that subsidies and price control would ensure that local tenants could afford to purchase houses in Docklands. The first schemes to go on sale seemed to support this view. However, as the number of house completions rose, the number of local purchasers decreased. This can hardly be surprising – local incomes are very low. While there was an initial unmet demand, this was soon fulfilled as those who wished to purchase, and could afford to, did so. In the first Barratt scheme in Beckton 13.5 per cent of houses went to council tenants. On the second scheme to be completed this dropped to 4.2 per cent, and by 1986 in a later phase of building no council tenants bought a house in the first release, despite the special nomination system.

The scale of the housing programme has taken everyone – even the LDDC – by surprise. Initial estimates were that the LDDC might double the population of 39,000, living in 14,000 households. Of the new housing about 10,000 units would be constructed on land owned by the LDDC, and a further 4,000 on privately owned land. However, this can now be seen to be a considerable underestimate. Despite the recession in house sales, a huge number of units have already been built – around 20,000, with sites for many more thousands identified, often with planning approval.

The massive housing programme means that Docklands has gone from a no-go area for developers, through an initial hesitancy and heavy subsidy by the LDDC to encourage construction, to the area with the greatest housing programme in London. It rapidly became fashionable to sport a luxury *pied-à-terre* in a converted Wapping warehouse – although apparently this is already no longer the case.

CONCLUSIONS

While in strictly statistical terms each borough has a more 'balanced' population profile, in reality it is an increasingly divided population. As more and more owner occupiers move in, those remaining in the council houses are made increasingly aware of the fact that they have little opportunity to change their situation. The government has no intention to help the councils; if anything their position has hardened, as was made crystal clear in the House of Commons in January 1986 by the then Under-Secretary of State for the Environment, Sir George Young, who said of Docklands 'it would be wrong for there to be further investment in public sector stock in that area'.

The LDDC was created with no specific 'goals' – except to 'regenerate' the Docklands area. Unlike local authorities it was not targeting local unemployment; indeed, it proudly announced many supposedly 'new' jobs which were nothing more than transfers to cheaper or newer premises from outside the Docklands boundary. Similarly, in terms of housing no specific goals were announced for numbers of new homes, or targets for rented or low-cost homes.

Far from 'priming the pump' for development, and then letting the private sector take over, the 'pump' has become progressively more difficult to move. It has needed more and more public subsidy to sustain the area, and yet developers such as Olympia & York, developing the largest property scheme seen in the UK, are still highly critical of the whole Docklands approach. Ironically perhaps, it would appear that developers *like* planning.

The community in Docklands has seen massive developments, but these have been of remarkably limited benefit to it. New housing for sale at high prices doesn't help. There are thousands of jobs in offices available just down the road (or the Docklands Light Railway) if a job in an office is what is wanted. Having more of those jobs brought right to your doorstep does little to make them more appropriate to a population with limited educational achievements, and few opportunities to obtain the training necessary to gain such jobs as do become available.

Moving the occupiers of the office space from the City or its fringes to a more remote location such as Canary Wharf seems to be of limited value to London as a whole. The occupier may pay a lower price for their accommodation (albeit mainly owing to the public subsidy provided by the Enterprise Zone regime). All that happens overall, though, is that office occupancy rates in the rest of London decrease. Meanwhile the manufacturing and service jobs which used to exist in Docklands have been moved on, or bought out, and luxury riverside homes replace the industries of the past few centuries, and the tens of thousands of jobs which they represented.

Docklands started as a story of hope; a dream of opening up the area to meet the needs and aspirations of the East-Enders who had lived there for generations. Once hi-jacked by the private sector developers in league with a new market-led government-sponsored approach, it rapidly turned into a nightmare of deregulated planning and massive over-development. The huge glass- and marble-clad offices have little of relevance for the local community, and represent a long-term monument to how 'regeneration' can become a disaster in less than a decade.

REFERENCES

Association of London Authorities (1991) *Ten Years of Docklands: How the Cake Was Cut*, London, ALA.
Broackes, N. (1981) *The Law Society's Gazette* 4 February.
Brownill, S. (1990) *Developing London's Dockland*, London, Paul Chapman.
Docklands Consultative Committee (1990) *The Docklands Experiment: A Critical Review of Eight Years of the London Docklands Development Corporation*, London, DCC.
Docklands Joint Committee (1976) *London Docklands Strategic Plan*, London, DJC.
Heseltine, M. (1980) Letter to Docklands Forum, 2 May.
LDDC (London Docklands Development Corporation) (1982–) *Annual Report and Accounts*, London, LDDC.

—— (1982a) Unpublished report to LDDC Board on public consultation and attitudes to the Corporation.

—— (1982b) 'Corporate plan', unpublished.

Ward, R. (1981) Talk at the Polytechnic of Central London, 10 December.

—— (1982) 'London Docklands, the LDDC's aims', *The Planner* July.

12

A MICROCOSM: REDEVELOPMENT PROPOSALS AT KING'S CROSS

Michael Edwards

They threatened its life with a railway-share;
They charmed it with smiles and soap.

> (Lewis Carroll, *The Hunting of the Snark*, 5)

The aim of this chapter is to elucidate a cameo of the physical and social development of London in the 1980s.

Just north of King's Cross and St Pancras stations in Central London lie about 40 ha of land, water and ill-maintained buildings representing a layering of transport history: the Grand Union Canal, the early stations of the Great Northern and Midland Railways, their goods depots, once one of London's main generators of horse-drawn traffic, the gasholders remaining from the Imperial Gas Light and Coke Company's works, stations of the world's first underground railway – the Metropolitan – and of four later deep 'tube' railways (Hunter and Thorne 1990).

Since August 1987 a debate has raged over proposals by British Rail (BR) and some lesser landowners for a massive office-led redevelopment of this under-used site. It lies just within the London Borough of Camden (LBC), and adjoins Islington (LBI).

The project became a partnership with developers Rosehaugh and Stanhope whose subsidiary, the London Regeneration Consortium (LRC), commissioned a 'master plan' from architects Foster Associates. The proposal is linked with a U-turn by BR who, in 1988, revealed their plans to bring TGV services through King's Cross, despite having told Parliament a few years earlier that this would be out of the question. An important thread in the whole story is the behaviour of BR and, behind it, the Department of Transport. People in most countries love to hate their state railways but in Britain it is hard even for an ardent believer in the potential of public enterprise like the present author to overlook the bumbling incompetence and deeply secretive habits of this body, especially visible in everything to do with the Channel Tunnel rail links. There is a bitter joke on the left which goes:

'The only good privatization would be BR.'

'Why?'

'Because it would be bought by the SNCF and run properly.

To some extent criticism of BR may be unfair; its inability to plan strategically, invest adequately or use its mass base of popular support has a history of many decades in which successive governments have bungled transport policy and starved the system. It is not the aim of the chapter to allocate blame for the situation.

There is strong local resistance to BR's plans for the LRC redevelopment and to the international station. Decisions on both are pending at the time of writing: by the planning authorities on the redevelopment and by Parliament on the station.

From the perspective of the promoters of the scheme there is simply a success story to tell about a lovingly designed redevelopment scheme which promises to bring derelict land into use and give London a fine new office centre linked with a station on the international TGV network, all with minimal public expenditure. The anxieties of local people and councils can readily and cheaply be allayed through some 'planning gain' deals. The central state, having set the market-oriented ground rules, keeps a benign distance and sees its strategy vindicated.

This perspective comes via a distorting mirror, however. This chapter argues that what we are seeing in the King's Cross proposal is the largest instance yet of a cumulative process which weakens the structural capacity of London and the human capacity of its people.

Long-term economic growth as well as environmental sustainability are threatened. This is not a case where the interests of capital are united, with government support, and opposed only by a rump of socialists, bent on redistribution to the poor. The state (central government and some elements in the local council) has used its high degree of autonomy from the interests of most forms of capital to ally itself with some very narrow interests: the only clear gainers from this project would be some developers and their backers, owners of nearby properties and shareholders in some recently or prospectively privatized corporations. Some of the prospective losers, however, are active and organized in resisting the plan and have had the confidence to come up with their own set of alternatives, developing their counter-proposals to a degree unprecedented in Mrs Thatcher's England.

At the time of writing London has a severe crisis in its highly volatile real estate markets so it seems unlikely that much new speculative office production will happen in the next few years. This gives us a break during which elements of a saner strategy are beginning to emerge. If MPs, local activists, local councillors and the more independent-minded planning officers can keep their nerve, better counsels could prevail before the next upturn sees construction under way again in London.

The slogan from Paris of 1968 applies: 'Do not adjust your vision; there is a

164

fault in reality'. But my argument is not an extremist one. The contradictions and (from almost everyone's point of view) incompetence of London's land-use and transport planning, of railway strategy and of labour-force policy generate raised eyebrows all over Europe. And although many European cities are struggling to generate economic growth without violence to their populations and sustainability – and no-one has 'the answer' – our particular mess is probably worse than you could find in Paris, Milan, Madrid or that great new investors' honeypot Berlin. I am not arguing merely that the incremental chaos of London is anathema to socialists, but that it constitutes a pretty self-destructive form of capitalism too.

These arguments are, so far, largely qualitative and, to varying degrees, tentative. The perspective set out here has been worked out by a small team at the Bartlett School of Architecture and Planning in University College London, working with local people but with minute financial resources.[1] Our main aim has been to provide some research support to local community groups, councils and the wider public but we have not been able to elaborate and press home these wider implications, or at least not yet. It is my hope that some of the hypotheses presented here will be picked up, developed and refined (or discredited) by others.

HOW DO WE THINK ABOUT IT: AUTONOMOUS PERSONALITIES OR PRISONERS OF THE SOCIAL STRUCTURE?

The analysis of social processes like the King's Cross development saga is always in danger of falling into traps. One trap is structural determinism: the presentation of events as the inevitable outcome of forces and contradictions in the social structure. The other traps is the opposite: analysing the actions of autonomous individuals and groups as free agents. Personalities and decisions are seen as the key issues.

In the first approach human beings seem powerless to determine or influence their future: they escape responsibility for what they do, since their freedom of manoeuvre is implicitly denied. The second approach tends to exaggerate the autonomy of people and groups through disregarding the constraints and imperatives which bear upon their actions. The social sciences have been trying to strike a balance where structural constraint and human agency each receive their due weight and it is this balance which is attempted here. (These problems of method are discussed by Ball *et al.* 1985; Chambert 1989; Jessop 1982; Page 1991; and in BISS annually.)

The chronological history is summarized in the *Narrative* box, and the text explores the agents involved and their relationships to each other.

The narrative

1987		Channel Tunnel Act passed
	spring	Rosehaugh Stanhope discussions with BR; DEGW pilot studies
	August	First leaks to press
	October	Land owners announce competition between 4 developers
1988	February	BR exhibits drawings by the 2 short-listed firms
	June	LRC/Foster selected
	September	Revised LRC plan
	November	LBC holds first King's Cross Exchange
		BR puts King's Cross Bill to Parliament
1989	April	LRC submits first application for outline planning permission to LBC
	July	LBC holds second King's Cross Exchange
		RLG publishes *People or Profit?*
	October	LRC submits modified plan
1990	April	Martin Clarke commissions KXT
	May	RLG Planning for Real starts
	July	House of Commons passes Bill on to Lords after long hearings and with modifications
		Third LRC outline planning application
	September	KXT 'Planning Weekend' UDAT held at UCL
	November	Expected submission of KXT plan to LBC deferred pending further technical work
1991	early	Rosehaugh and Stanhope declare financial problems; sell some assets to reduce borrowing
	February	LPAC advises LBC that LRC application could not be determined until BR proposals settled
	June	LBC's expected decision again deferred, pending negotiations. Strong pressure on Council to declare themselves 'minded to approve'
	July	Rosehaugh and Stanhope merger talks announced

THE PLAYERS

The social relations of modern Britain, and the specific forms they take in London, can be seen in the agents, groupings and individuals active in this story. Some like British Rail (BR) are hard to classify between capital and the state but that is part of the story: London sees a blurring of the boundary between public and private spheres and the penetration of private accounting and accumulation criteria in to what were previously public bodies.

166

Investors in real estate . . .

. . . provide much of the dynamism of London's booms and slumps. They are thus a sensible starting-point. A peculiarity of building production in the UK (discussed in the conclusions of this book) is that there is only the most indirect relationship between demand and supply. It is rare for people or firms who need buildings to commission them direct. Most buildings are, instead, produced speculatively, much like cars, in the hope that someone will buy them. Unlike cars, though, most users of commercial buildings rent rather than buy. It is as though Hertz, Budget and Godfrey Davis were the main customers for cars. Their equivalents in commercial building have tended to be big financial institutions – insurance companies, pensions funds and, recently, banks. They don't usually get involved in development until construction starts, but they are increasingly dominant from then on, especially over the market in completed buildings.

So far, at King's Cross, we don't have financial institutions since construction is not yet under way. The developers will, however, have had discussions with potential financiers and their responses will have had some effect on the proposals. The developers' strong links with Olympia & York and with Japanese and other banks are discussed below.

The importance of these investors has been that it was *their* confidence in the continuing growth of returns on Central London property which drove capital values of land and buildings to their 1988/9 peak. We got to the point where prime buildings were changing hands at such high prices that current rents only yielded 5 per cent returns on investment. Such a market is a highly fragile affair, depending entirely on continuing scarcity and confidence. It can, and did, quickly topple into a downward spiral.

Landowners

The speculative office boom in London encountered a grave shortage of building sites when it began in 1984–5, especially of central sites suitable for 'large floor-plate' buildings. These buildings, with large areas on each floor, were then thought essential to meet the needs of the financial services conglomerates which were forming in preparation for the 'Big Bang' deregulation of the securities markets. Developers and some of the key opinion-formers who advised them (e.g. Duffy and Henney 1989) laid great stress on the technical obsolescence of pre-1970s buildings. This 'technological forecasting' was not entirely well grounded. The assumption was that computers would proliferate, produce more waste heat and need more cables, that air conditioning would be universally required and that financial dealing, at least, would need huge open floors. It was already clear, however, that waste-heat generation per unit of computer power was declining, that optical fibres would simplify cabling problems and that air conditioning would carry substantial health and comfort penalties. And the demand for huge 'dealing rooms' was certainly over-stated: the number of dealing firms has declined since the crash of October 1987 and it never was clear why screen-based traders needed uninterrupted views of each other. Was it to permit supervision on the lines of Jeremy Bentham's 'panopticon' prisons or was it just pandering to dealers' nostalgia for the days of blackboards and waving arms?

The Docklands, of course, was available and offered plentiful sites, many of them enormous. But the failure of the state adequately to invest, in advance, in the rapid links to the City and to housing areas which would have valorized these sites meant that they were deeply unappealing to most investors.

To some extent the chase after sites, and even after large sites, was met by the relaxation of planning controls in the City of London – the 'square mile' of the financial district. In 1987 the Corporation of the City of London adopted a new plan, its first since it was liberated from the influence of the GLC. This new plan sought to give fresh impetus to the City's pre-eminence through three main relaxations: (1) plot ratios over wide areas of the City were enormously increased;[2] (2) conservation area status was withdrawn from some zones and (3) space below ground was excluded from the calculation of plot ratio. These measures (combined with the extension of value added tax to building restoration) triggered a substantial round of redevelopment in the City, intensifying a zone which was already the greatest concentration of office work in Europe.

But the Docklands and the City did not absorb the whole boom; outlying areas of the centre had two strong attractions for developers: large sites, notably of disused railway land, were being brought to market there. And, for those developers willing to go a few hundred metres away from the conventional 'prime' areas, the gains were potentially enormous.

Office rents were highest round the Bank of England – so high that construction costs could be covered by just two years' rent in the mid-1980s. Furthermore these rent streams were discounted by investors at very low rates. The main beneficiaries of this situation in the 'prime' zones were the owners of the land and property since investors and developers were competing strongly with each other for the opportunity to build or rebuild to satisfy this ultra-cautious demand for 'prime' buildings. But actually the rents paid by occupiers were rising fastest on the periphery of the centre, a few hundred up to a few thousand metres away from the peak (see Figure 12.1). There were thus golden opportunities for developers and investors innovative enough to take marginal sites and transform them into high-prestige quarters.

British Rail, the main landowner at King's Cross, is an instance of an essentially 1980s phenomenon; a state agency increasingly deprived of state funding but still prohibited from raising money on the private markets, under increasingly strong imperatives to make profits from each and every one of its assets. Such agencies were, during the 1980s, being privatized or prepared for privatization, which meant that they increasingly adopted private-sector accountancy practices, basing their investment and operating decisions on short-run financial criteria. BR had been a deeply inactive landowner but now became hyperactive in the

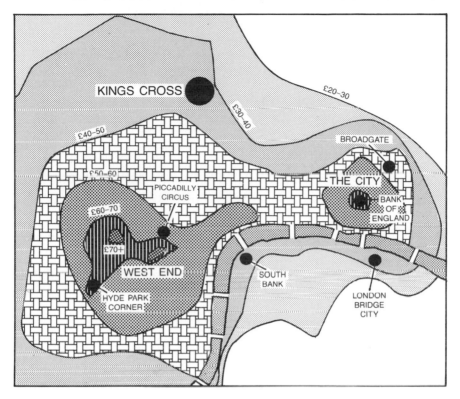

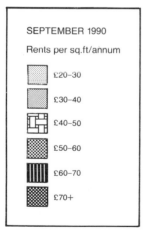

Figure 12.1 Location of King's Cross with Central London office rent contours

Source: Hillier Parker.

valorization of its Central London stations and attached land.

At King's Cross BR had inherited, at nationalization, the land rights of the railway companies – rights which Parliament had transferred to the Great Northern and Midland Railways in the nineteenth century.[3]

The rest of the 40 hectares or so of land was in the ownership of other public bodies and a few small private owners. The most important, after BR, was British Road Services, privatized as the National Freight Consortium in a buy-out by management and workers earlier in the 1980s. This organization owns a compact but crucially important central block of ex-railway land, partly covered with listed buildings and mostly occupied in 1988 by its own operating subsidiaries (in trucking) or by tenants on very short-notice leases. Like so many of the privatized agencies of the 1980s it inherited extensive land and buildings which, in the lax and 'flexible' planning environment of the period it was able to valorize through disposal or development.[4] By the late 1980s NFC was preparing itself for the extension of its shareholder base from the workforce to the general market so it was an organization very actively concerned to have a healthy balance sheet. It was thus structurally placed to join a landowners' consortium, led by BR, with alacrity. But NFC went one step further in that it also joined with developers Rosehaugh and Stanhope to constitute a development consortium for the site.

The crucial rôle of the landowners in this saga was that, in 1987/8, they organized a competition among developers for the rights to the project. The details of the competition have never been made public but we do know that a short list of four developers was narrowed to two. Each of the two then produced architectural proposals (which were displayed in public) and financial bids (which were not).

The landowners' choice was based on these design-plus-money offers. Clearly the landowners formed the expectation that they could make a killing. It is understood that, if the project went ahead on the scale envisaged by the developers, the landowners would receive a long-term equity share of 70 per cent of the surplus profits, and that an advance payment of £400m. against this stream would be made at the outset of the project.

In the author's judgement, this 'auction' of the development rights, in a planning environment where the extent of these 'rights' was not pinned down precisely by zoning plans, has created the situation we are now in: the developers can say that they have very little choice but to build a massive commercial project because it was on that basis that they estimated their bid for the site.

It is in this way that 'reality' gets defined. When the local campaigners propose an alternative which they judge viable it can be dismissed as 'unrealistic' by LRC and BR because it does not measure up to *their* agreed version of reality.

The developers

After the crash of the early 1970s it was widely considered that development companies as an autonomous force in the UK would be largely subordinated by pension funds and insurance companies (the 'institutions'). Some were but, as Carragher (1986) demonstrated, the property and development companies were a diverse bunch, some financially very strong, rich in cash revenues and well able to remain autonomous. Furthermore the institutions have spent the 1980s slimming down the relative weight of property in their portfolios and have sometimes even been net sellers of real estate. They have been busy fuelling the equities boom. The 1980s property boom was fuelled from two main sources: the internal resources of the cash-rich property companies and investment by banks. Banks doubled their net investment in property companies between 1983 and 1987 while the institutions were halving theirs. The banks concerned were UK and overseas banks in varying proportions, with Japanese, Scandinavian, Swiss, EC and North Americans very active.

The lean years of the early 1980s saw some remarkable innovations in development financing by a handful of new entrants to the development sector. Rosehaugh and Stanhope, the victors of the competition at King's Cross, were of this group, as was the under-bidder, Trevor Osborne's Speyhawk.

These 'new' developers were masterly at marshalling short- and medium-term syndicated bank loans on a huge scale. They found ways of raising these loans through partly-owned subsidiary companies whose debts they did not have to declare in their own (parent companies') balance sheets. This device, known as 'off-balance-sheet funding' protected the assets of the parent company since the loans were secured against the land, bricks and mortar of each project, without recourse to the corporate wealth of the parent. The same device also minimized the apparent gearing (debt ratio) of the parent company. In a period of rapid rental growth and stable borrowing costs this was a viable – indeed brilliant – procedure. But of course it is highly vulnerable to falling rents, rising interest rates or combinations of the two (Wong 1988).

But it would be wrong to attribute the rise of Rosehaugh and Stanhope simply to financing systems. They have been great innovators in other respects. In architecture and building they have superseded the old production system whereby architects are commissioned to make a design, the design is put out to tender and the main contractor is then left to build. Stanhope in particular have reorganized the whole process, following American practice to a great extent and frequently using American architects (e.g. SOM for Broadgate) and construction managers (e.g. Schal alongside Bovis). They follow a system of commissioning and building geared to a regime of significant real interest rates (you build quickly with costly borrowed money), the need to catch a volatile market (especially, at Broadgate, the demand generated by the 'Big Bang') and great dependence on 'value engineering'. Value engineering is the process of being as scientific as possible in spending money on quality where it contributes to profit but minimizing spending elsewhere.

These new practices are thus a design and building process in the office sector analogous (in its profit-maximizing effectiveness) to the operations of Britain's speculative house builders, well understood through Ball's analyses (1983).

Rosehaugh and Stanhope gained much of their standing through financial and technical innovation. But a third novelty has been the public images of their respective chairmen. Godfrey Bradman, of Rosehaugh, is rather 'green', active in support of Friends of the Earth and a pioneer in specifying environmentally friendly materials for buildings. Stuart Lipton, his counterpart at Stanhope, is equally a challenge to the stereotype of the philistine developer. He's very active in the arts and in university governance and has sponsored a number of seminars on the social impacts of development. David Dickenson, one of his co-directors until 1991, is active in public debates on metropolitan planning – one of the prominent voices calling for a revival of strategic planning – and chair of the East London Technical Education College.

The complex character of these contemporary development firms is not widely enough understood. Local councillors and MPs concerned with King's Cross, for example, make regular speeches about the 'greed of developers'. But in my view there is nothing particularly greedy about the developers at King's Cross. They are driven by nothing more than capitalism's 'normal' pursuit of profit ('normal greed', if you like), just as if they made computers, cars or clothing. Indeed they go about their business with an energy and creativity which, in Amstrad, Fiat or Benetton, is hailed as economic innovation. It must be galling for them to be so reviled. And it is naïve of those who call them names. If we find their project socially unacceptable it must be a general objection to what gets produced under capitalism – and perhaps a criticism of the particular framework within which they operate in Britain. Public planning and policy here presents developers with opportunities like King's Cross and the behaviour of public and private landowners ensures that competition between developers generates massive development proposals.

So in 1987 we find these two companies, the Davids, not the Goliaths of property development, teaming up with NFC to form a development consortium for the site, winning the auction and starting a long struggle for permission to build (see Figure 12.2).

The local authorities

We now know that Rosehaugh Stanhope were starting to set up the King's Cross project by at least early 1987, less than a year after the destruction of the GLC. It is likely that the GLC's approach to strategic planning would have reduced the chances of a major permission at King's Cross. And there had been specific barriers to major development there under the GLC's Community Areas Policy (discussed in George Nicholson's chapter, above). The government, however, 'deemed these policies to have been withdrawn' when the GLC was abolished in 1986. A very specific hurdle was thus removed.

The Borough of Camden thus became the front-line authority, responsible for granting or refusing consent. Its Borough Plan was up-to-date (London Borough of Camden 1987), but of the kind which is stronger as a qualitative statement of

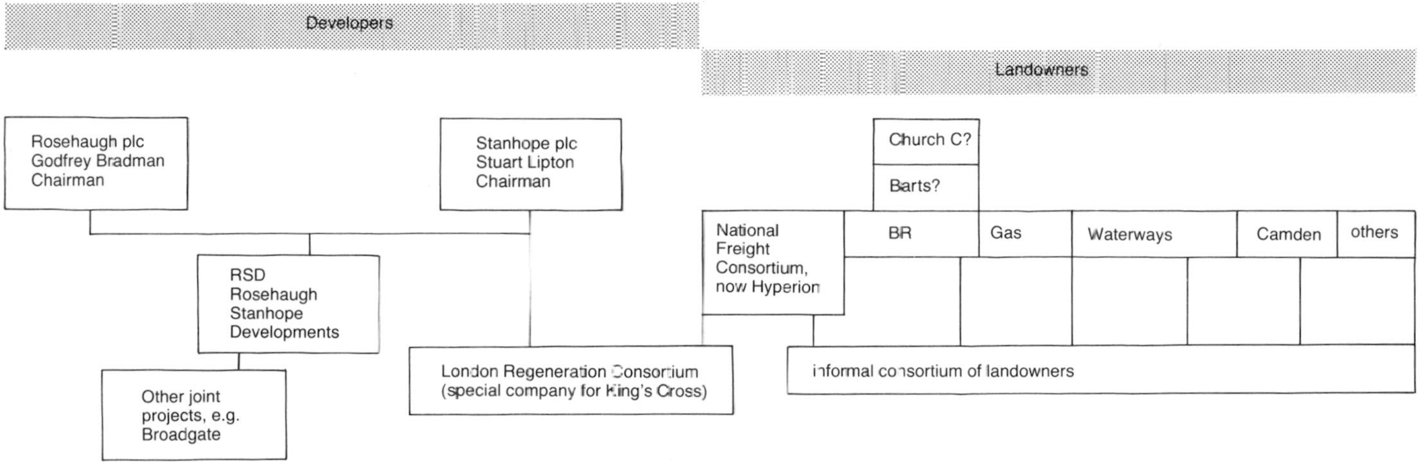

Figure 12.2 King's Cross agencies

intentions than as a rigid yardstick of what is permissible. When it discovered Rosehaugh Stanhope's intentions it lost no time in preparing a 'Community Planning Brief' (London Borough of Camden 1988) which amplified the provisions of the plan. This document, and the Camden plan behind it, constitute a curious mixture of excellent technical work and political irrelevance. Its strengths are in enumerating what would be desirable in a King's Cross redevelopment: better transport interchange, community facilities for a very deprived area, affordable social housing, access for disabled people, and so on. Although it did set targets for housing numbers and the 'social' housing content, its weakness was that it did not offer a binding quantification of the maximum employment growth the transport system could accommodate. These things were essentially left for negotiation. Camden was thus, in my view, largely at the mercy of the developers' version of 'viability'.

It is also important that Camden and its neighbour Islington were both suffering from the severe fiscal stress placed by the Thatcher regime on left-leaning London boroughs. Their staff resources and capacity to propose active strategies were thus severely stretched. They did manage to assemble a small planning team and do some technical work. They also mounted two substantial and well-executed 'planning exchanges' – big public participation meetings each lasting through a weekend – and mobilized revenues from parking charges to help fund the work.

A major consideration in Camden's behaviour, however, has been to try to avoid a public inquiry into the LRC scheme. This could arise either through a refusal, followed by LRC appealing, or through the Secretary of State calling the decision in for his own determination. Camden judged that they could probably secure a better scheme through negotiation than they could hope to get out of an inquiry followed by a decision of a Minister hostile to their views. They also have in mind the risk that, if they were judged to have refused permission 'unreasonably', the costs of an inquiry might be awarded against them – a potentially devastating penalty for a cash-starved borough. It does appear that the developers have put enormous effort into persuading Camden that alternatives, of the kind they and the community groups want, would not be 'realistic', so the councillors will have to be very tenacious and well-informed if they are to refuse permission.

The developers' motives for avoiding an inquiry are less clear. Three factors are probably involved: avoiding the money costs of a major inquiry, avoiding delay and maintaining an image of being community-friendly, rather than confrontational and rapacious. Thus we have seen a curious kind of consensus of developer and council both trying to agree in order to avoid an inquiry.

Two other local public bodies are involved. Islington Borough, which adjoins the LRC site, contains many houses and businesses at risk from BR's international station redevelopment and from the rippling property price rises which would flow from LRC's scheme – which indeed are already taking effect. Islington is not directly party to the decision and at times has seemed unlikely to oppose LRC because it hoped to gain some material compensations. At the time of writing, however, it seems to be more resolute in opposing LRC on traffic, housing and broad planning grounds.

We also have the London Planning Advisory Committee, a consensus-oriented statutory committee of the 33 London Boroughs with a staff of 16, the nearest we have to a strategic planning agency for London. LPAC has become very concerned about the cumulative effect of redevelopment in Central London in expanding employment beyond the capacity of the transport network and of the housing stock (Swain 1990). It is also strongly committed to supporting growth in East London and thus has strong reservations about the BR proposal to site an international station at King's Cross. It has been advising Camden in 1991 in quite negative terms about the LRC scheme.

Central government

Spatial planning in the UK has long been characterized by a high degree of contradiction and disharmony between ministries. Policy and action on land use, transport, industrial growth and labour markets and training fall to four minis-tries. This dislocation seems almost as endemic in the thinking of the Labour opposition as in the Tory governments of the 1980s. For King's Cross this is sig-nificant in explaining the failure of co-ordination of London public transport capacity and investment with need and demand. Transport policy in London (see Chapter 4) reached the end of the 1980s in particular disarray. The road propo-sals in the Department of Transport's London Assessment Studies (which could, among other things, have brought a major highway access into King's Cross) roused such opposition that they were unceremoniously dropped. And uncer-tainty over routes of international trains caused a similar furore. The two débâ-cles combined to unite disparate community groups against the government to a degree not seen in the decade.

In contrast, the DoE's impact on King's Cross has been reticent and obscure. Its general stance has tended to favour commercial development proposals against resistance from boroughs, and on the LRC scheme it has, publicly, merely reserved its right to call in the case for decision if it chooses to do so.

The other key rôle of the central state has been over the planning of the TGV network and stations, DTp insisting that railway route decisions are a commer-cial matter for BR to decide, and Parliament still being the arbiter where BR seeks powers to do railway works. BR has a Bill before Parliament to give it power to build the King's Cross international terminal. A further BR Bill to permit links through London to King's Cross has not yet even been laid before Parliament. Skilful and superbly orchestrated local resistance has delayed the King's Cross Bill in the Commons and severely weakened it,[5] and at the time of writing it is working its way through the Lords. Parliament may yet see that it is quite irra-tional to permit a station before considering the network which would serve it; so, on these or other grounds, the Bill could yet be thrown out. Equally Camden could decide that it cannot logically determine the LRC application until the railway routes are approved.

Professionals

The choice of professionals to advise LRC has been highly significant. Instead of asking an experienced planning firm to prepare what must be Britain's largest city-centre redevelopment proposal, LRC appointed architects: Skidmore, Owings & Merrill (SOM) of Chicago and subsequently Foster Associates. The professions of architecture and planning have diverged so far in the UK that a highly problematic vacuum exists in between. It may well be true that few UK planning firms have the vision as well as the technical capacity to grasp such a vast project. Foster's office certainly has vision – and is the source of perhaps our finest architectural modernism – but it equally certainly lacks experience and skills in planning. But it was a shrewd choice, given that much of the battle over King's Cross was to be over capturing media attention, winning the hearts and minds of journalists, cameramen, residents, councillors and governments.

The LRC 'Master Plan' prepared by Foster Associates, with a large supporting cast of specialist consultants, is in its own way a superb response to the brief (Figure 12.3). The objections are essentially to the brief and to aspects of the design (bulk, height, social and traffic impacts) which are inevitable consequences of trying to fit 700,000–900,000 square metres of building on the site. Given that requirement, the intention that most of the building should be corporate offices and the early decision to make a central oval park the key 'image', there are relatively few options. Building height over much of the site is constrained by the need to protect views from Hampstead to St Paul's, so the only possible designs are for very deep, solid slabs of building. Later versions of the plan have slightly reduced the height over most of the site to meet local resistance, but reinstated most of the lost floorspace in two tall towers in the northeast corner – a decision which seems to arouse as much local resistance as did the previous versions, especially in Islington, over which long afternoon shadows would be cast.

A distinctive feature of the scheme is the extent and high cost of decking over railway tracks. This has a dual explanation. In part it results from the admirable effort to integrate the new development physically with the surrounding social housing areas by having a lot of direct links. Fosters have been strongly and effectively advised on this by Professor Bill Hillier and a team at University College London, and the scheme incurs high costs to avoid being a gigantic cul-de-sac, surrounded by tracks. The second reason for decking is simply to fit the target floorspace on the site. The argument here is partly a circular one; this quantity of commercial floorspace is needed to make the scheme profitable, even on pessimistic assumptions and after covering the high costs of paying in the early years for the decking as well as the land price.

Popular movements

The LRC site lies just outside the northern edge of what is usually thought of as London's metropolitan centre. The area has a strong history of socialist local government and very large-scale council and housing association building of working-class flats. Thus within about 1,000 metres lived about 60,000 people in 1981, nearly two-thirds of whom were tenants in social housing estates. The population is highly diverse, with a strong survival of trade union and traditional Labour Party organization. Overlaid on this is a diverse range of professionals, skilled and unskilled manual workers, students, unemployed and retired people. Ethnic diversity is high by London standards, but without very strong concentrations of particular groups. Levels of housing stress, unemployment and poverty are high by London standards, but not the worst to be found. A very wide range of local organizations exists: tenants' associations, resident, amenity and conservation groups, transport campaigns, a group of Indian businesses, industrial employers on the LRC site, church groups, women's organizations, and so on. Both Camden and Islington have somewhat decentralized administrations and the (Labour-dominated) political life of the area is active and has a local area, as well as a borough-wide, focus.

A number of local activists took an initiative in summer 1987, when rumours of the redevelopment first leaked out, in forming an 'umbrella' campaign which came to be known as the Railway Lands Group (RLG). This group, and some of its more autonomous component groups like Crossfire in Islington, have proved to be highly effective in articulating resistance to the LRC scheme and to the BR international station proposal – though interestingly not to the basic concept of redevelopment, which commands wide support.

There is no space here to recount the long saga of this struggle but some key features must be pointed out. A distinctive feature of good campaigning in the 1980s has been getting the right target. In the 1970s most struggles directed themselves at the state, usually at local government, as the agent able to decide events. Now the capacity of councils to determine events is in some doubt, and their fiscal capacity to back decisions with actions that require money is in tatters. Campaigners have rightly addressed themselves simultaneously in three directions: at the councils and councillors certainly, but equally at the developers and landowners directly and at the media. The RLG's capacity to capture press and TV coverage has at times been superb – and crucial in challenging the LRC version of reality. In Mrs Thatcher's London we began to see oppressed and frustrated people confronting capital head-on, with less mediation by the state than before.

A central problem for all campaign groups, however, is money, and the administrative and professional work it can buy. The RLG is no exception in being chronically starved of resources. With great difficulty it gained some tens of thousands of pounds from the Councils, some from the RIBA Community Architecture fund, and helped secure the £16,000 of research support which has

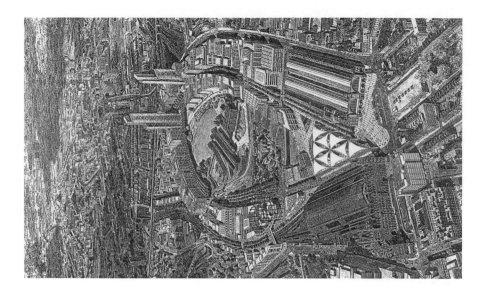

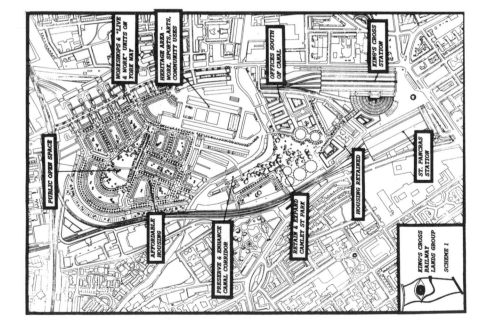

WORKSHOPS & "LIVE & WORK" UNITS ON YORK WAY

HERITAGE AREA – WORK, SPORTS, ARTS, COMMUNITY USES

OFFICES SOUTH OF CANAL

KING'S CROSS STATION

PUBLIC OPEN SPACE

HOUSING RETAINED

ST. PANCRAS STATION

AFFORDABLE HOUSING

PRESERVE & ENHANCE CANAL CORRIDOR

RETAIN & EXPAND CAMLEY ST PARK

KING'S CROSS RAILWAY LANDS GROUP

SCHEME 1

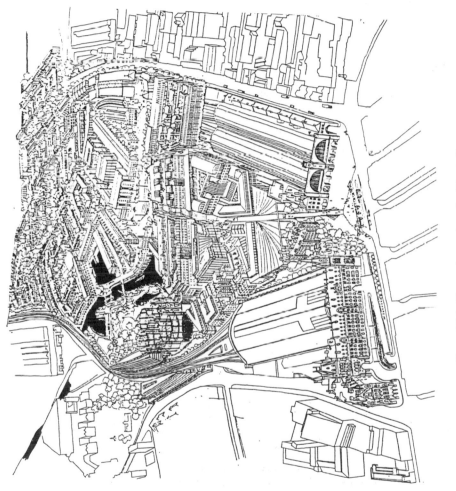

Left: KXT	
offices	374
private housing	75
social housing	149
total building	712

Top left: RLG	
offices	22
private housing	51
social housing	154
total building	350

Above: LRC	
offices	545
private housing	101?
social housing	50?
total building	785

All figures '000 m^2

Figure 12.3 Alternative projects for King's Cross

enabled our team at University College London to provide some technical advice (Edwards *et al.* 1990).

In 1990 a local surveyor and developer, Martin Clarke, became involved, investing over £100,000 in technical studies for an alternative plan. This plan was prepared under architect/planners Norman Sheppard and Ian Haywood, constituted as the King's Cross Team (KXT). It sought a compromise which would carry strong community support while being financially 'viable', albeit at a much lower level of payment for the land. The central target has been to reduce the office content from LRC's 700,000 square metres to below 400,000 and to increase the housing content, especially the social housing, to something approaching Camden's and the RLG's targets. The work was consolidated in a 'planning weekend' in September 1990, organized by the architect John Thompson and modelled on earlier Urban Development Action Team interventions (UDATs). This extremely stimulating event brought together professionals, local activists and residents in a five-day brainstorm and design project which generated a draft plan, report and public presentation (Cowan 1990). Further elaboration of the KXT plan has followed (see Figure 12.3) and further funding is now being assembled to get it to the stage where planning permission could be sought. The RLG has greeted the scheme as far preferable to LRC's, but has refused to endorse it unconditionally. Indeed the RLG seems to have gained confidence that alternatives are really possible, and that further moves could be made away from the 'office city' towards meeting more social needs and goals and having a much greater diversity of activities.[6]

The collapse of the London office market since 1989/90 has further emphasized the high risks of pinning the viability of a development purely to this most volatile of speculative building markets. It does not make sense from a local labour market point of view either. If jobs in corporate offices were going to make significant inroads into Inner London's unemployment levels we would have seen dramatic improvements in Camden and Islington in the late 1980s. As the last major study of London employment found, efforts to slow the displacement or demise of existing jobs probably do more to reduce unemployment than do the new jobs we generate (Buck *et al.* 1986:193).

Martin Clarke simultaneously funded a large-scale 'planning for real' exercise, led by an RLG steering group and managed professionally by Michael Parkes, a planner and veteran of Spitalfields. This exercise has probably been one of the most ambitious constructive consultation/participation exercises yet mounted, drawing on hundreds of hours of meetings, manipulation of huge site models, outreach visits to enlist non-joiners and public debates. The outputs have been a People's Brief report (Railway London Group 1991) and a further set of alternative plans, prepared by Michael Parkes, Daniel Mouawad and Mark Scott under the RLG's constant guidance (RLG forthcoming).

CONCLUSIONS

Fundamentally the importance of the King's Cross story is the struggle over whose definition of reality is to prevail. In Mrs Thatcher's decade it came to seem normal, natural, even essential for progress, that the LRC/Foster plan should go ahead. It seizes with both hands the economic and design opportunity presented by the TGV interchange; it brings derelict land back into use; it dramatically extends and reshapes London's central business district; it offers a fine park, lined with prestigious addresses for European corporations. Its main street, with trams down the centre and lined with trees, brings the sophistication of Zurich's Bahnhofstrasse (with offices over shops) to a run-down part of London and ties the orbital North London Line railway in with the tube system. It maximizes the physical integration of the area with its surroundings and tries to avoid dead areas.

It is a compelling vision, and its power and presentation have beguiled many. People from all over the world arrive in London expecting to find it built or under construction. Even the shrewd urban analyst Paul Cheshire has written (1990) arguing that the public decision on this scheme is a litmus test of whether London is serious about maintaining its world-city rôle.

Yet all this is a profound mistake. Putting a major international station in London's nineteenth-century heart is to miss the opportunity it presents. Imagine putting a substantial airport there. To get the economic benefits of such a (potential) growth machine one would put it where there is plenty of room for road access as well as public transport, and for the snowball of development, jobs and incomes which could follow it. Stratford seems a better site. The Royal Docks might be better still. But no systematic comparative analysis has been done of the effects on London's structural development.

Putting this station, with a major corporate office centre, in the middle of London in an intricate, congested, mixed-class housing area is to ensure the threat of displacement and gentrification for tens of thousands of people and many firms. King's Cross is, granted, the best-connected point in the London Underground network but there is little or no spare capacity in that system. Do we really want to commit the huge resources necessary to create the capacity there? This author, at least, would expect a serious study of the alternatives to recommend a more decentralized structure.

Added impetus to this view would come from serious attention to the energy and pollution consequences of further employment growth in Central London. Rapid cumulative growth of demand for professional and clerical labour has tended to be met not locally but by growth in long-distance commuting. Is that the future we want?

SANITY TO THE RESCUE?
POSTSCRIPT, OCTOBER 1991

The summer and early autumn of 1991 have brought two significant shifts.

At the level of London planning, the Secretary of State for Environment became persuaded of the merits of the East Thames Corridor Strategy which had been initiated by SERPLAN and supported by LPAC. This plan aims to co-ordinate infrastructure and new urbanization along the north and south banks of the estuary. An integral feature is to bring the Channel Tunnel Rail Link (CTRL) in to London from the east, rather than from the south. This change of heart reached the newspapers in August, attributed to the influence of Mr Heseltine's special advisor, Professor Peter Hall, who had been making frequent trips back from his job in California to advise at Marsham Street.

Then at the Tory Party conference in October it was announced that Mr Heseltine was to commission urgent studies on the corridor strategy, while the Secretary of State for Transport, Mr Rifkind, had decided to scrap BR's proposed route from Ashford via Swanley and a deep tunnel to King's Cross and commit the government to the alternative via Rainham Marshes and Stratford which had been proposed by engineers Ove Arup and backed by the Manufacturers' Hanover Trust.

This decision on the route is described as 'final'. But a long delay in starting work is envisaged, so many things could change in the meantime. There is still no suggestion that nation-wide networking of high-speed trains is being planned. Questions also arise over that element of the Arup scheme which links Stratford with King's Cross in a bored tunnel, popping up on the same alignment as the BR route and thus able to use the 8-platform sub-surface station below the King's Cross train shed. The implausibility of this link is that it envisages the costly drilling of a pair of tunnels from Stratford in to Central London only a few years after another pair will have been completed to take the Crossrail to Paddington via Tottenham Court Road. Are we really to believe that public or private money would be found for two such projects? The argument could only be sustained with overwhelming evidence that this would be the only way to secure good through-train services to the north and west. The evidence on this remains thin and the proponents of Stratford are busy refining their counter-claims for northward routes from there.

The second policy shift has come from Brussels. On 17 October 1991 the British government received a letter from Carlo Ripa de Meana, Environment Commissioner, calling for work to stop on a number of UK projects, including the Ashford–London railway and the King's Cross redevelopment, on the grounds that they are in breach of the EC environmental assessment directive. The press reports that his objection is due partly to the fact that the environmental appraisals (and the respective decisions) have been separated into three: the link, the new station, the commercial redevelopment, and partly to the likely substantive impact of the terminal's location at King's Cross. The RLG is celebrating, and perhaps their lobbying in Brussels has helped secure this intervention. The LRC and BR are keeping brave faces and insisting that nothing has changed. At the time of writing those elements in the Camden planning office and committee who have been working for a permission for LRC seem to be holding their line.

In these new circumstances the RLG has resolved to submit its alternative schemes as planning applications to Camden to re-emphasize the real presence of alternatives and as the focus for discussions with agencies in the public, private and voluntary sectors who could be members of development agencies or trusts. And the plans themselves have been given the second of two prizes for innovation awarded by the RTPI and RIBA for the London region.

NOTES

1 Sincere thanks to the London Boroughs of Camden and Islington and to the London Strategic Policy Committee who between them subscribed £16,000 to support the work. The views expressed here are the author's, not necessarily those of the sponsors, and draw on the contributions made by team members Ellen Leopold, Mike Geddes, Daniel Mouawad.

2 The plot ratio is, roughly, the ratio of maximum permitted floorspace to the ground area of the site. It is the main mechanism for regulating the density of commercial building.

3 BR's position has been complicated by the discovery that some of its acres had been acquired by the railways under parliamentary powers which endured only so long as the land was needed for railway purposes. Should the railway use cease, the land would revert, at the original price, to the previous owners – the Church Commissioners and the Trustees of St Bartholemew's Hospital. This ownership dispute remains unresolved.

4 NFC's land activities are now in the hands of a subsidiary, Hyperion Properties.

5 A key modification has been the removal of a section which would have given BR a special right to circumvent the normal powers of English Heritage in respect of Listed Buildings.

6 Another major boost to RLG's confidence had been the 1989 publication of a popular account of the finances of LRC's plan (RLG 1989) and BBC TV's coverage of this report.

REFERENCES

Ball, M. J. (1983) *Housing Policy and Economic Power*, London, Methuen.

Ball, M. J., Bentivegna, V., Edwards, M. and Folin, M. (1985) *Housing and Urban Planning: A European Perspective*, London, Croom Helm.

BISS (annual) *Proceedings* of the Bartlett International Summer School on the Production of the Built Environment, London, University College London.

Buck, M., Gordon, I., Young, K. with Ermisch and Mills (1986) *The London Employment Problem*, Oxford, Clarendon Press and Economic and Social Research Council.

Carragher, M. J. (1986) 'Property development and accumulation', unpublished M.Phil. thesis, University of London.

Chambert, H. (1989) *Long Term Structural Changes in the Building Industry: Sweden 1950–85*, Stockholm, Nordplan.

Cheshire, P. (1990) 'The outlook for development in London', *Land Development Studies* 7:41–54.

Cowan, R. (1990) 'New challenge on King's Cross', *Architects Journal* 26 September, 14–15.

Duffy, F. and Henney, A. (1989) *The Changing City*, London, Bulstrode Press.

Edwards, M., Leopold, E. and Geddes, M. N. (1990) *King's Cross Railway Lands: Second Report*, London, Bartlett School, University College London (research commissioned by London Boroughs of Camden and Islington and by the Railway Lands Community Development Group).

GLC (1984) *Alterations to the GLDP*, London, GLC.

Hunter, R. and Thorne, R. (eds) (1990) *Change at King's Cross from 1800 to the Present*, London, Historical Publications.

Jessop, B. (1982) *The Capitalist State*, New York, New York University Press.

London Borough of Camden (1987) *Camden Borough Plan*, London, LBC.

—— (1988) *King's Cross Community Planning Brief*, London, LBC.

Page, M. G. (1991) 'The local state and housing production programmes: a study of change in the 1980s and the case of Haringey council in North London', unpublished Ph.D. thesis, University of London.

Railway Lands Group (1989) *People or Profit?*, London, RLG.

—— (1991) *A People's Brief*, London, RLG.

—— (forthcoming) *King's Cross Railway Lands: Towards a People's Plan*, London, RLG.

Swain, C. (1990) 'Technical evaluation of large-scale schemes and cumulative impacts', proceedings of a seminar *Managing Urban Change: Strategic Impacts*, London, Stanhope Properties plc, 9–17.

Wong, B. K. H. (1988) 'Recent changes in commercial property finance', unpublished M.Sc. thesis, University of London.

13

A VISION FOR LONDON

Michael Edwards, Kelvin MacDonald, Peter Newman and
Andy Thornley

INTRODUCTION

Previous chapters have shown the many dimensions of London's crisis. However, not least of London's problems is the crisis of ideas on how to move forward. It is widely accepted that pragmatism, muddle and the market are not up to the job. The received wisdom of the moment is that London needs a vision. For example in 1991–2 a consortium of voluntary groups has been organizing a Vision for London festival. Meanwhile Local Authorities in London, through the vehicle of the London Planning Advisory Committee (LPAC), have also been exploring ways of progressing a London-wide perspective in the wake of the abolition of the GLC. This crystallized in LPAC's *Strategic Planning Advice for London: Policies for the 1990s* (1988) and its co-sponsorship of the World Cities Project (Coopers & Lybrand Deloitte 1991). The Labour Party also produced its document *London: A World Class Capital* in 1991. As we shall see the debate has been taken up by the private sector. This widespread interest immediately raises the question: whose vision?

One vision is market-led – the pathway to innovation, flexibility and speed. It is pointed out that the future quality of life for Londoners is based upon economic growth. Thus the private sector must be allowed to flourish, particularly in the context of competition from other European cities. Such a view leads to structures of strategic decision-making in London which are free from the inertia of democracy and parochialism. The Urban Development Corporation presents a model of this approach.

A strategic body would co-ordinate infrastructure, provide a framework for private investment and develop a vision around which resources and promotion can concentrate. There is much talk of 'leadership' or 'a voice for London' whether of French Presidential variety or American City mayor; civic pride is back on the agenda. In Victorian times this provided the impetus for moderating the desolation of Britain's cities caused by the industrial revolution. Cities competed with each other for the most pleasant image – new parks, public facilities and services and grand cathedral-like town halls. Now we have the competitions for city culture, garden festivals, new museums and tourist attractions. Everyone is marketing an image. This is taken right down to details: city crests on manhole covers, hanging baskets and uniquely designed street furniture.

There is no doubt that London has a lot going for it in relation to culture; however, it does have one disadvantage in the post-modern society of image promotion. What is London? Who identifies with London as a whole? London has always been described as a collection of villages. The centre of the city is dominated by a transient population, whether of office workers or tourists. This problem is not helped by the organization of local government. These bodies do not relate to communities, thus creating problems for participation, local identity and civic pride. What sense of common feeling exists between Dulwich and Brixton, between Muswell Hill and Tottenham? However, the citizens of London must be involved in formulating and realizing its future, otherwise any vision will dissipate in the face of people's alienation and disillusion. Riots do not provide the right kind of image. Schemes which lack consent are often stopped politically or at least badly delayed.

Thus there are two dilemmas that have to be faced in considering the future of London. First, how can the varied vested interests and differing needs of Londoners be brought together into a vision which can act as a vehicle for co-ordination and planning? Second, how can this be achieved in a way that incorporates local democracy and community involvement in the face of imperatives of international capitalism?

Any vision and mechanism for change has to develop out of the present. There is a danger that a concentration on the development of visions, although essential as a vehicle for galvanizing support and moving forward, ends up as a free-floating capsule disconnected from reality. It is important therefore to have a good understanding of the current forces at work as a basis for building any vision. As King (1990) has amply demonstrated, London operates in the context of international capitalism which generates many forces outside the control of national government let alone local government. On the other hand, we have seen in earlier chapters that many aspects of London's crisis have been caused or, at the very least, enhanced by central government's ideology and policies over the last decade. This interplay of international, central state and local arenas requires a detailed and thorough analysis which is beyond the scope of this book. However, we feel we cannot move to a discussion of new approaches for London without some references to the forces that operate in the capital.

ELEMENTS OF A DIAGNOSIS

The great paradox of capitalism is that it has an up-side and a down-side. It gives us (or some of us) growth, innovation and change simultaneously with exploitation, misery and a deep conservatism about things that matter. London's experience in recent decades has been just such a mixture – and the Thatcher years have been especially extreme. The social effects of Thatcherism have borne hard on Londoners as we have seen in this book – in deteriorating services, housing stress, polarization of incomes, and so on.

For the economy as a whole we see Thatcherism as having been a dead end. Admittedly, after great misery in the early 1980s it produced a spurt of growth in the later 1980s. But its approach was to increase the incomes of investors by smashing trade unions, making labour relatively cheap, lowering taxes on firms and on the incomes of their proprietors. The argument was that this would make the economy 'leaner and fitter' in some sustainable way. In fact, when the history comes to be written, we shall see that we have fallen between two stools. We don't have the capital-intensive, research-based, production which has been the secret of German growth: success through rising productivity. Nor do we have cheap enough labour (yet) to play a rôle like that of Portugal, Poland or Ireland, let alone to compete with South-East Asia. The weakness of this real economy makes it hard to preserve the world status of sterling as a currency. But the City of London is very influential and the world rôle of the currency has had to be protected at all costs. The cost *is* high: everyone in the economy has been required to live with an over-valued currency and high real interest rates: we can see the effects in our housing costs, in the fiscal problems of local councils, in the impoverishment of public services and in the inability of manufacturers to invest. It has been a weakening, not a strengthening, of the economy. London has suffered this weakening very acutely. As we saw in Chapter 3, it lost its industry rapidly in the 1970s and 1980s and much of its replacement activity has been carried on the back of booms in the financial services and property development sectors. But these have been fragile booms and they have not lasted.

More and more aspects of life have come under the market 'discipline' of Thatcherism. Privatized utilities like water and gas start trying to realize speculative property gains by selling or developing their reservoirs and gas works. Services which remain in the public sector experience the same pressures: health services, education authorities, universities and councils compete to sell 'surplus' land and buildings in a desperate attempt to balance their books. British Rail, the proprietor of large stations in Central London, makes millions from developments above and around its termini – though at the price of distorting its investment planning and in flagrant disregard of the planning consequences.

In this situation there was great growth, dynamism and innovation in the commercial property markets where great speculative profits stood to be made. But this has been problematic too. The UK is characterized by a strongly developed dualism in the commercial property markets: there is an *investment market* where buildings are bought and sold by corporate bodies interested in the flows of prospective rent and/or capital appreciation. The buildings are then rented out to users (retail, industrial and service-sector firms and public bodies). Buildings are thus produced primarily to meet the needs of the investment market, rather than directly to meet the needs of building users expressed in the *user market*.

The significance of this dual market is not yet adequately understood at national or local level, though the phenomenon is widely discussed (Nabarro 1990; Ball 1985; Luithlen 1986). Clearly, though, in London it fuels the volatility of the boom-and-bust cycles. The property markets simply are not self-stabilizing: they behave more like farm products, where glut years alternate with shortage years. It also affects the work of architects and building workers – and the environment which is produced. So we get increasingly uniform speculative buildings, targeted at an abstract developer's concept of the prime tenant. The work of architects is reduced to maximizing square feet and rental value and adding enough cosmetic appeal to secure planning permission – often by an emphasis on real or imitation historical detailing.

The investment and development markets of the 1970s were dominated by the UK pension funds and life insurance companies, but their influence has waned as they have reduced the relative weight of real estate assets in their portfolios during the 1980s, and in some years have even been net sellers. Their place has substantially been taken by banks, British and overseas, who have been substantial net lenders to property companies. The Bank of England estimates that outstanding property-sector debt to the banks had exceeded £40bn by 1991.

These investors are fundamentally financial capital, moving their funds in search of accumulation in its value (Harvey 1982). The property markets are thus fragile markets dependent on the maintenance of expectations of constant future growth. In London, as in Japan and the USA, we are finding out how severe the crash can be when such booms end. And, since public-sector building activity has dwindled, the construction sector has no work to fall back on when the commercial market collapses.

Similar problems would be found through analysing the housing and retail markets. However, are all the problems due to these market processes? Our argument is that they result from the way markets inherently tend to work in the absence of good planning systems. Our planning systems have been peculiarly bad at getting the best from markets and avoiding the damage. Partly this reflects government policy – the abolition of strategic planning, the disaster of the LDDC, the deregulation of use classes – and partly from governments' failure to understand what is happening. The extreme flexibility and vagueness of the land-use regulation system greatly increases the uncertainties confronting developers. They don't know what, if anything, will get them a permission, and they don't know how much competition there will be from other sites. So on the whole they choose safe locations and build conventional buildings. Perhaps most seriously, they can no longer rely on state provision of infrastructure, or on the state to regulate the quantity of development to match what the infrastructure can stand. The 1980s have certainly shown us many things which markets cannot do: to orchestrate transport networks and buildings is clearly one of them.

THE PURSUIT OF DIVERGENT INTERESTS

One underlying feature of London's crisis has been the way in which different interests have sought to pursue their own needs, according to their own imperatives, with little contact between each other. In rather oversimplified terms, one can say there have been those interests which are concerned about the future of

London as a viable business environment and those interests which are concerned about the future of London as a place to live. The first group is highly sensitive to London's competitiveness and focuses on wealth creation and London's image.

Such a perspective has, on the whole, tended to ignore the needs of the less well off, relying on the principle of 'spin-off' to automatically benefit everyone. This neglect has been given legitimacy by the Thatcherist ideology and its acceptance of inequality in society. Meanwhile the various citizen groups who have been trying to represent the needs of their members have found themselves increasingly isolated from decision-making. Local authorities have found themselves stuck in an unenviable position in the middle. On the one hand they are drawn into making their locality more economically attractive, supporting the aspirations of the first group of interests, while also being subjected to political pressure from the citizens of their area. Bound by central government policies – in particular, financial constraints – they have found themselves unable to counter the trend towards greater dissatisfaction and alienation of citizen groups. What strategies are the two broad categories of interests pursuing to try to meet their objectives?

Growth coalitions

The development of what has been termed 'growth coalitions' has been a feature of the inter-city competition in the US (for a review of the literature on this see Leitner 1990). These coalitions are made up of business élites, unions, media and academics and involve the co-option of local authorities. The approach is for the business élites to form a particular organization which then prepares an economic strategy for the city. The media is used to disseminate the ideas and mobilize popular support, while the academic element of the alliance gives greater credibility. This propaganda dimension is used to promote the idea that the strategy will benefit everyone.

The state is brought in to provide the necessary means of achieving the strategy: infrastructure, finance to reduce risks, relaxation of planning controls, compulsory purchase, and so on. As a result, in the US there has been a shift in local state expenditure from the provision of services and public goods to initiatives that are orientated towards trying to promote economic growth, e.g. through loans, tax incentives, land purchase and improvement subsidies. This trend has been accompanied by a more entrepreneurial style of government and a change in attitude by politicians who are faced with ever increasing fiscal problems and an inability to meet the needs of citizens. They are therefore attracted towards ideas that might seem to provide a solution to the cities' predicament and show that they are doing something positive. In the inter-city competitive climate they feel they need to help make their city attractive to investors (Harvey 1985). They can present themselves as guardians of citizens against the rest of the world.

This strategy can be criticized for by-passing local democracy. The instigation of public policy by business élites is 'sold' to the general public through the media. Politicians are then obliged to go along with the strategy because of their financial impotence. Local democratic representation does not feature in this approach, a situation which is accentuated by the range of new undemocratic agencies that are usually created to implement the strategy. As a result the voice of many groups and interests in the city's population is excluded. Once again there is the reliance on 'spin-off' from economic growth to eventually meet the needs of such groups. This demonstrates a faith which we do not share.

However, perhaps these trends only apply to the US and will not spread to Britain or, in particular, London? Lloyd and Newlands (1990) have described how this growth coalition idea has already reached our shores. Aberdeen was in economic difficulty as the North Sea oil boom came to an end. In response to this, local business interests formulated a new economic growth strategy called 'Aberdeen beyond 2000'. This strategy supported some aspects of existing plans, especially the road-building programme, but contained many elements that contravened the prevailing policies: for example, redevelopment of historical areas, lifting restrictions on office and shop development in the central area, de-regulating green belt, and closing schools and releasing land resources for development. The group said that the local council should 'put sensible use of resources before local political issues' (quoted by Lloyd and Newlands 1990:54). Following the growth coalition approach they sought to gain public support through the use of the media and directly appealing to the citizens of Aberdeen (a summary of the strategy was posted through every letter-box).

Could this happen in London? Certainly the constant reference to the external threat of European competition has set the climate for such a coalition. Meanwhile the LDDC has set the precedent for the non-democratic agencies which are an important element in the implementation of growth coalition strategies. There are other signs of moves in this direction. The CBI (1991) has proposed a London Development Agency which would comprise a small government-appointed group to promote the city and attract investment. At the time of writing it appears that the government are considering such an agency, if not for the whole of London then perhaps for the East Thames corridor. The developers Stanhope have hosted seminars on London's future, drawing in key opinion-formers, especially business and development interests and academics. The property consultancies firm of Hillier Parker have also proposed an appointed strategic body to include business interests (Robinson 1990). The embryo of a growth coalition for London is therefore in place and ready to develop. Before discussing further the relevance of such an approach for London, let us turn to the other broad category of interests, i.e. those directly related to the needs of citizen groups in London.

The citizen perspective

There is no single 'citizen perspective'. This has to be built up from the views of the variety of different groups in the population. As discussed in the chapter by Nicholson there will be communities who experience common needs based upon the characteristics of their locality. However, as shown in the chapters by Valentine and Cross, there will be other groups who share experiences of deprivation which are not necessarily tied to where they happen to live. These different dimensions have to be built into any future vision.

The view of many citizen groups is that they face bureaucratic, political and financial barriers in their attempt to change the places in which they live or work. In local communities and other interest groups there are very clear ideas about what needs to be done. Groups are frustrated in their attempts to get somebody to do it, or to get into a position where they, themselves, can take on the job of improving the quality of their lives. The failure both of government to respond positively to citizen demands and of the market to provide for local needs has given rise to the rapid growth of the third sector of voluntary, community and specific interest group associations. There has been a switch from protest against the flood of office blocks to setting up local agencies which can deliver the affordable housing and local jobs needed in the inner city. There is now a huge variety of community companies, businesses, co-operatives and development trusts. All of these initiatives need to be built into a new vision. The answer to the question of how London should develop is in many cases already there; the problem is that citizen groups are having trouble implementing their ideas.

There are some success stories. For example, the Coin Street sites on the South Bank are owned by a community company which plans and manages one of the biggest mixed development schemes in London (Brindley et al. 1989; Tuckett 1990). Involvement in such schemes offers direct participation in the planning of London but this has not been achieved without difficulty. The first problem and one that is common to almost all community projects is money. Local communities usually want to build affordable housing, provide community services, or set up enterprises which banks and other money-lenders regard as unprofitable. Capital funding of community schemes is a major barrier to a citizen-based vision for London.

Money is not the whole of the answer. Communities in Central London have always found public sector investors less willing to put up money for their projects because high land values make the costs much higher in the centre than in the suburbs. The Housing Corporation said that it could get better value for money by building in Hastings! The concentration of financial power in Central London means that land values are higher there than anywhere else, and indeed only higher in New York and Tokyo. It is not the fault of Londoners that their land is expensive. They don't all want to move to Hastings, but it makes little sense to buy up land at hugely inflated values. This problem should be solved by using the planning system. But unlike many other countries we do not have a clear system of land-use zoning which restricts international office blocks, and the high land values they produce, to a planned area. A flexible system encourages land speculation; every landowner wants the highest price. A clearer system of plans which meant what they said would allow lower-value housing land to be allocated in Central London. A further simple change is needed to planning law which would enable plans to distinguish between social housing and housing at market prices.

Any plan for London needs to identify land for local communities. At present the presumption in favour of market-led developments means that social housing and facilities are often only available through bargaining with the developers. Developers and communities do not always have common interests! More clarity in the planning system would overcome the endless conflicts over local land-use issues. The Labour Party (1990) looked at adopting ideas from France which would allow the zoning of special areas where land for housing and other facilities could be bought cheaply, and at giving local councils the right to first refusal on any piece of land which comes up for sale. Speculative land markets cannot exist readily in the French system. However, this is not preventing Paris from competing with London as the international headquarters of the new Europe.

A citizen-led plan for London needs a supply of money and some basic legislative change. It also needs a change of attitude in government at all levels. All community groups have heard officials say that what they want can't be done or that they are being unrealistic. Some local councils have become more responsive to direct citizen involvement, and decentralization and the encouragement of tenant control indicate a movement towards greater participatory democracy. The local politician can no longer be the sole representative voice of a ward or constituency and the official should no longer claim a monopoly of knowledge of what's best for local communities. Building on the achievements and ideas of citizen groups offers the opportunity of a future made up of many different types of initiative and ways of doing things, for experiment and for learning across London. There will be no single answer to the variety of demands.

We need to dispel two myths about community participation. The first is that local communities and interest groups are always parochial. Under the last Labour GLC a large number of community groups took part in the review of the strategic plan for London and indicated their ability to think on a London-wide level about the future. There has always been a strong tradition in London of groups getting together to fight road proposals, revived again recently in response to consultants' reports on opportunities for new toll roads in London. In Chapter 9 we saw how groups threatened with similar pressures contributed towards a broader strategy of dealing with office development leading to the Community Areas Policy. The second myth is that communities will always be negative; will only be interested in stopping a proposal. The story of King's Cross in Chapter 12 shows that if groups are given the opportunity they can become involved in proposing plans and solutions. This demonstrates the point that it is time, resources and opportunity that prevent a more positive approach – not lack of interest. So what government structures are needed for the development of a new vision?

DEVELOPING DEMOCRATIC STRUCTURES

A positive role for central government

Planning

On several occasions in this book caution is advanced against popular images of Paris as a city better governed and better planned than London. We can, however, use the French experience to clarify some of the issues facing London.

In both cases national government has been the determining force behind recent change. In Chapter 10 Hebbert argues that the fate of London needs to be set in a national policy context, a context of diminishing public services and inadequate investment in infrastructure. In Paris the major infrastructure developments – the TGV network, the Grands Projets, expansion of the Universities, expansion of the office quarter, investment in social housing – are all determined by the President and the Prime Minister. The first lesson from Paris is that the government of capital cities cannot be left entirely to the markets and local powers. However, it comes as no surprise that local political resistance to heavy involvement of central government in the affairs of London is echoed in Paris. The current process of updating the regional plan has made political conflicts plain. One of the reasons for the peremptory withdrawal from discussion of the regional plan was that the regional council is controlled by the right and the government is socialist. At present, the government's draft plan published in April 1991 sits alongside the plan of the regional political leaders. The future of both awaits regional elections in 1992. Notwithstanding these kinds of conflict it would be naïve to propose a future system of government and planning for London without some central government intervention. Central government must take up the challenge.

Finance

Part of the GLC's case against its abolition was that it redistributed resources around the capital. The richer boroughs effectively paid for services and projects in poorer areas. Since 1966 this redistribution has been centralized. Government decides between the needs of different boroughs, and in the case of politically favoured boroughs such as Westminster and Wandsworth, central government has been able to offer local citizens a very low Poll Tax. The redistribution of resources across the Paris region is called for in the Socialist Party's ideas for the new regional plan. The French Government introduced legislation in the spring of 1991 to change the formulas which allocate resources between municipalities.

Councils controlled by the right – such as Paris – complained as strongly as do Labour London boroughs about the inequity of the allocations and the threat to local autonomy. The lesson from Paris is again that we should expect central government to set resource priorities across London.

There will be a continuing political debate about who should gain or lose in the allocation of resources. Hebbert criticizes central government for not investing in London's infrastructure and for a lack of co-ordination. In terms of administrative structures Paris has the advantage of a government system which is co-ordinated both horizontally between central departments and vertically through the system of prefects. The 'problem' in the government of London in relation to Paris is not central versus local powers but of poorly co-ordinated and informed central planning mechanisms. Reform of the structure of central government is a priority.

Local government: from representative to active democracy

It is not only Paris which has an elected mayor. The system is widespread in western democracies. The United States offers a more attractive model to the Conservative Party because the political management structures are more stream-lined. Cities have fewer councillors; they do not always represent local wards but have a general rôle. Executive power is not exercised through the 'cumbersome internal management' of the committee system but by a mayor or elected manager.

What can be learned from mayoral systems depends on one's political point of view. An executive mayor can seem more businesslike or offer the possibility of a strong voice speaking up on behalf of cities. Direct election brings with it the personalized politics of presidential campaigns. The lessons from the United States, however, suggest caution in the hopes for progressive reform. We have seen that London is becoming a more polarized city. City politics in US cities seems to reflect a similar polarization. Mayors may support programmes to relieve poverty, housing programmes, and education and training for minorities. But such programmes attract little private or central government support. Policies which prioritize local needs, and better management in service delivery – the sort of localist strategies that some London boroughs have adopted – run into funding crisis. Sooner or later the land and property market imposes choices on localism. The dominant form of local politics in US cities is one of collaboration with the private sector. The city mayor for the most part is not the representative of the 'underclass' but a partner in the growth coalitions we have described above. Mayors can represent citizens, but also narrower economic interests and national political interests. The new city hall in Tokyo cost £500m. and repre-sents the construction companies which built it and the power of conservative party politics.

Since the 1987 general election Labour London boroughs have pursued increasingly localist strategies. Many boroughs share the common desire to prove

194

to both central government and local electors that they can manage their local communities efficiently and effectively. New concerns with quality have been spurred on by the threat of compulsory competitive tendering and privatization of local services. Decentralization and devolution are widely seen as offering a desirable future for the management of local government. In some cases decision-making has been transferred beyond local council offices, and housing co-oper-atives have shown the way towards 'user control' of services. It is the level of politics below the boroughs that is crucial to the quality of democratic life in London. We have argued that central government must take a more inter-ventionist role in the future of London. A much stronger local politics building on organized citizen groups is needed as a counterweight. How can community-based politics be incorporated into government structures? Local government has been changing, and the diverse lessons from attempts to create a more caring and 'enabling' local government need to be shared. It is the job of the town halls to facilitate citizen activity. This means recognizing the limits of representative democracy.

Greater flexibility in the government of London is a welcome side-effect of the abolition of the GLC. Hebbert points out that the boroughs have managed to form all sorts of joint bodies to deal with common interests. As long as these bodies are democratic a multi-agency political arena offers more opportunities for citizen voices to be heard. Unfortunately at present both main parties seem to favour a unitary local government system. The quality of political life is dimin-ished by having one standard local government authority.

London is planned by a non-interventionist government and weak local councils trying to manage private interests. The alternative is for strong central direction counterbalanced by a multi-agency, open, conflict-ridden, challenging, local democracy (with or without the voice of a mayor).

The planning of Paris is a closed debate between political élites, planning professionals and civil servants. In New York development decisions are managed by the mayor and developers. Planning in Britain has a unique history of participation, and local government has adopted more participatory struc-tures. The future must develop these innovations to produce democratically discussed solutions. Greater flexibility, an enlarged 'not for profit' sector and more active local democracy will produce diversity. The increasingly polarized city needs as many diverse responses as it can get.

TOWARDS A VISION

Thus one of the principles is full involvement in the development of a new vision. It is therefore inappropriate for us to set out what we think this vision should be. However, we can suggest some aspects that we believe are important and raise some ideas for discussion. It is time to bring together the three elements that are needed if any solution is to be attempted – new visions, new structures and new actions. These are intertwined but can all too easily be mistaken one for another.

It is not enough to have any one without the others. Nor does the existence of one lead to the existence of the others. A new London-wide body does not in itself create a vision; an action such as a more effective drive against litter does not demand a vision; and the espousal of a vision certainly does not mean that anything at all will happen. It is for this reason that this book has moved freely between these three components. In bringing these elements together, however, we need to focus first on visions.

It should not be necessary, at this late stage in the book, to have to argue the need for a 'vision' for London to be clearly and publicly articulated.

Now, the articulation of a vision is not the same as a plan for London. The government has stated (DoE 1989) that 'London's future depends on the initiative and energy of the private sector and individual citizens and effective co-operation between the public and private sectors, not on the imposition of a master plan'. Such a statement may be anathema to traditional land-use planners, who once thought that the future of London lay in their hands, but may be welcomed by those who failed to see any sense of vision in the 1976 Greater London Development Plan. In this, the government are at one with the thinking of the London Labour Party, which stated in the 1981 GLC election manifesto: 'London has suffered from unwieldy planning procedures, often based on unrealistic, utopian plans for the future.... The GLC ... will not commence any large-scale plan-making exercises'. Whilst this may sit uneasily with the subsequent work on the updated Greater London Development Plan (GLC 1984) the sentiment is still one to be applauded. London does not need plans – it needs planning.

Having made this distinction we can start with visions because the creation of a vision may, surprisingly, be the easiest of elements to agree upon. Anyone who knows anything of London may write their vision, and there is little doubt that certain key concepts will occur again and again. Most visions would include the desire for more ease of movement, a cleaner place, freedom from fear, an inspiring environment, access to jobs, more open and recreation areas, a choice in housing, greater fairness ... The words may differ but the underlying desires will remain the same.

One problem is that many of the better-known visions of London have remained at this broad level of abstraction. The current 'official' vision is contained in the Department of the Environment's *Strategic Guidance for London* (DoE 1989):

> London in the 1990s must be a city where enterprise and local community
> life can flourish, where prosperity and investment will continue to increase,
> where areas which had declined will find new rôles, where movement will
> become easier and where the environment will be protected and improved.

It is not clear whether the 'must' in this sentence is to be read as an imperative or as a wearied plea.

Compare this with the best-known unofficial vision – that of Peter Hall in *London 2001*:

What would such an organically planned city look like? How would it feel to live and work in? We can pick out four main elements. For work and also for services and entertainment, it would be a *many-centred city*. For living including community services and the education of children, there would be *real communities*. For moving about, we would have a *choice of transport systems from any A to any B*. For recreation or the plain enjoy-ment of looking, there would be a *continuous green backcloth*. And, behind them all, *a slimline regional planning system*.

(Hall 1989)

The emphases are the author's own. It may be unfair to criticize such summar-izing sentences for lack of content but, even taken with the fuller contents of the documents, they evade the central issues.

As we have already said, it is not the rôle of this book to put forward its own vision – this runs the risk of narrowing the perspective, as Peter Hall did when he described the future city through the eyes of one affluent, employed family – the Dumills – who could not hope to express how the city would work for those against whom it discriminated. We can however, start to build a set of criteria against which visions can be assessed.

First, it is apparent that the vision must show how it is to be achieved, it must contain the seeds of its own implementation. In this we move into the other elements of structures and action; but if a vision stays at the level of rhetoric then it fails in its rôle as the promoter of change.

Second, it must extend beyond the narrow view of planning as related only to the use of land or the appearance of place. A vision must deal with equitable distribution – whether this is of travel, of access to recreation, of finance, of housing, of pleasing places or of any other factors that can make the same physical environment a paradise for one but a hell for another. It must deal with the multitude of discriminations that prevent all Londoners from participating in all the city has to offer. One of the elements of the London Planning Advisory Committee's 'Fourfold Vision' is London as a City of Opportunities for All – 'to ensure that development policies reflect the needs of all Londoners and reflect a priority to the most disadvantaged individuals, communities and neighbour-hoods' (London Planning Advisory Committee 1988). Perhaps the greatest short-coming of the Department of the Environment's formal *Strategic Guidance*, which was meant to be informed by LPAC's *Strategic Planning Advice*, was that it failed entirely to reflect the polarity that exists in London.

Next, any vision must contain a breadth of concern sufficient to encompass the problems to be tackled. This needs to be looked at in three ways. First, one can start to debate 'What is London?'. It is certain that the old GLC boundary does not contain all the areas that need to be considered in an overall strategy – notably the area stretching east to the Thames estuary. On the same basis, one needs to explore the South-West, the East and West Midlands, and East Anglia if the majority of the influences on the regional economy are to be included. How far, however, do we go? One of Peter Hall's Dumill family commuted to Lille. On this basis one could stretch the boundaries of London ever outward and still not achieve a better set of solutions to the problems that face the capital. Perhaps, instead, there is the need to reduce the area of search so that any vision for London only encompasses that area that is truly London whilst trying to take into account wider influences. London is a unique place – it is not the same as the South-East and any strategy must encapsulate and build upon its unique nature.

Second, a breadth of vision can, and should, be taken to draw in a European dimension. The race is on amongst Western European capitals first to define their rôle and then to compete for whatever economic activity supports and feeds that rôle. For any city to compete in the new single market and in the world beyond, it must examine the way it looks, what it provides for residents and visitors and how it transports the people who use it. Allied to this is the fact that the European Commission is taking a more active rôle in urban and land-use policy. This movement has small beginnings. The first meeting of European Ministers of Regional Policy and Planning took place in November 1989 with a follow-up meeting one year later. In June 1990, the Commission published a *Green Paper on the Urban Environment* and, less than a year later, *Europe 2000*, which forms a preliminary view of the factors that individual countries need to take into account when preparing settlement plans. The Commission have already stated that they will take into account local plans when deciding whether to grant aid projects (Commission of the European Communities 1990) and have pointed out how many of their policies in other areas, such as environmental quality and regional development, impinge on national decision-making about urban areas (Commission of the European Communities 1991). As the EC moves more into this policy area, it is apparent that London needs pro-active strategies against which they can judge the policies of the Commission.

Third, breadth of vision encompasses more than just a width of geographical boundary. It includes, importantly, a willingness to stretch the imagination as far as identification of problems and ways of tackling them are concerned. There is a rôle for views of the future that look only at surface appearance. Judy Hillman's *A New Look for London* (Hillman 1988) does not pretend to do other than this and is no less useful because of it. Such an approach did, however, allow the then Prime Minister, Margaret Thatcher, to write a foreword which described the breadth of London's rôle: 'London stands at the heart of this great nation, and its well-being is vital to us all. It is the hub of our political, financial and business life, a major centre for the arts, shopping and sport, home to nearly seven million people ...'. But it then allowed her to display a paucity of vision in describing solutions, when she welcomed 'the emphasis on the need for everyone to involve themselves in the upkeep and improvement of their local streets'. The irony, plain to the reader, was lost on the author.

We have seen that a vision for London may be easy to write. The extreme difficulty comes in turning that vision into action. One of the keys to this is often conveniently forgotten. In order for any change to take place there must be a willingness for that change to happen. In 1944, reconstruction and the ending of squalor was a priority and Abercrombie's Greater London Plan could serve as a well worked out checklist for action. Such a unanimity of purpose has not existed since. However, the differing voices of the various parts of the 'community' of London have started to come closer together. Such willingness for change must not be stifled by whatever structures are set up.

Whilst designing the right policy-forming, decision-making and implementing structure for London is almost impossible, there are lots of ideas for the form that such a structure should take. We have seen how the development industry sees the need for an overall strategy for London and, therefore, for a body to carry this out. We have already noted the proposals of the CBI, Hillier Parker and Stanhope Properties. Other models are emerging in the form of community-led development initiatives, for example Community Development Trusts (Bailey 1990; Department of the Environment 1988). The Labour Party has also put forward its ideas for a strategic body for London – the Greater London Authority (Labour Party 1991). It is said that this will be a 'lean' organization. At the time of writing, too few details have been released to answer the key questions of implementation powers, funding or relationship to the London boroughs.

As shown in Chapter 11, the London Docklands Development Corporation has the power to act and the level of resourcing from central government that allowed it to do so. It fulfilled its function, however, in a way that has not satisfied even those who wished to prove that market forces could create workable places without bureaucratic controls. At the time of writing this book, however, there are indications that yet another agency is being considered for the regeneration of an area of London – in this case the East Thames Corridor. It is not clear whether this agency will take the form of another development corporation or whether it is to be modelled on another body, Scottish Enterprise.

This latter body has progressively moved into a position where some local authorities view it as being some form of powerful, government-funded super-developer whilst the development industry compains that it introduces an unhealthy distortion into the development market. Whether, therefore, any new body for the East Thames Corridor is set up as a development corporation or as a development agency along the lines of current models, development will take place. It is not certain, however, that the development will either serve the immediate needs of this part of London, or that it will provide a sustainable basis for its long-term development.

Whatever structures are adopted to implement a vision for London they must incorporate accountability. Many of the proposals outlined above do not adequately do this. We have already suggested that there is a need for both a strong rôle by central government and a counterbalancing vital and diversified range of local democratic agencies. However, there is a need for much more thought to be given to how these different tiers can be combined and to the rôle of local authorities and a London-wide body in the overall system.

So what sort of London might emerge from this interplay of markets and democratic practices? In a sense we cannot say. The whole point is that expert blueprints and rigid utopias are not the way forward. But we can say something about the main issues we might expect to see on the London planning agenda as the century comes to a close. We would expect the agenda to take new directions and adopt new approaches as it responds to the principle of putting people first: for example, a more humane city would result from listening to the voice of London's women. Recently London has lacked the framework and institutions to address such issues about the future. This was illustrated at a recent seminar on planning for the new Berlin (Ryser 1991). Planner Dieter Frick said that one of the reasons for coming to London was that they were wondering where to site the station or stations for international fast trains. He said that 'in London you have some experience of analysing where to put these stations to get the greatest benefits for the structural evolution of the city and we want to learn from you'. The British audience burst out laughing. After all we have spent years avoiding this constructive way of thinking about the issue. It illustrates the way in which the dominance of a market philosophy can narrow one's vision and destroy the forums that are necessary to debate alternatives.

A new agenda would accept the inadequacies of the market in meeting the needs of people living in London. Trickle-down does not work. To accept this opens up new opportunities: for example, it can be argued that an efficient transport system for a city the size of London can only be created with state support. Once this is accepted, then quality and the needs of all the different sections of the population can be placed on the agenda and, as we saw in the chapter by Bashall and Smith, all kinds of new initiatives are possible. Similarly in relation to housing, Brownill and Sharp have shown that a reliance on owner occupation has failed to meet the needs of many Londoners and a new, more innovative agenda is required.

In Chapter 3 we saw that the economy of London has been undergoing radical change and an agenda for the future has to respond to these changes while also meeting the needs of all Londoners. The future of the economy will always involve an interplay of priorities between nation, region and locality. From the localities (and from London employers) come desperate pleas for order-of-magnitude improvements in education as well as training. If the full potentialities of all Londoners are to be harnessed we shall need forms of education which release people from the overlapping discriminations of gender, ethnicity and the archaic class structure. The revitalization of London's high streets can be combined with the potential of the cultural industries to provide both new job opportunities and a more pleasant environment to live in (Comedia 1991; Montgomery 1990).

As argued by McLaren in Chapter 5, the new agenda will also have to give greater priority to the environment. It has to be accepted that if London were left to market forces it would run the risk of environmental collapse. There is a need for a new approach which looks at the environmental capacity of the city and its sustainability: a comprehensive and co-ordinated approach which considers the spatial distribution of activities and transport from the point of view of resource efficiency and quality of living environment. Punter has shown in Chapter 6 that there is a lot that needs to be done to create a good urban environment in London and that new ways have to be found to ensure that this reflects the needs and aspirations of Londoners.

These then are just some of the issues that we might expect to feature strongly on a new agenda. As we have already said the agenda would be generated and developed through the more open democratic structures we have suggested. However, could all this talk of constructing new agendas and new perspectives be criticized as head-in-the-sand utopianism? Don't we really have to accept that market forces should be given their head if London is to compete with other cities in global competition? Our answer to this is that, at the end of the day, any economic system is built upon the actions of people. Any economic system is also vulnerable to crisis and can create conditions which are detrimental to its own development. These adverse effects include the antagonism of the people on whom the system depends as well as a disregard for environmental damage which can, in the longer run, create problems for continued economic progress.

Back in the early nineteenth century the British economy suffered probably its greatest crisis in adjusting to a new stage of capitalism. This was accompanied by appalling environmental conditions in our cities and popular insurrections on a massive scale. The resolution to the crisis was achieved through the acceptance of the interrelationship between economic development, the living conditions of people and the environmental quality of cities. This was epitomized in the manifesto which Disraeli presented in his novel *Sybil* – the solution took the form of an enlightened industrialist who cared for the life of his workforce outside the factory and constructed a village for them of high environmental quality. The economic benefits of this approach were illustrated by the example of Sir Titus Salt, who was so inspired by the novel that he constructed such a village. He later claimed that the village was the best economic investment he ever made. The economy is once again going through a transformation creating many problems and, as we have shown, London is in crisis as a result. Now the message of history is that the way forward out of this crisis is once again to acknowledge the interrelationship between economic progress, people and environment. Obviously we do not want the paternalism of the nineteenth century, and people must be involved in decision-making. A resolution to London's predicament has to come by widening the agenda and approaching the issues from new perspectives. The vision must include the views and needs of London's citizens in all their variety and take a long-term and comprehensive approach to the environment. London's economic health in the future depends upon the vitality and happiness of its citizens and a pleasant and sustainable physical environment.

REFERENCES

Bailey, N. (1990) 'Community Development Trusts – a radical third way', in J. Montgomery and A. Thornley (eds) *Radical Planning Initiatives*, Aldershot, Gower.

Ball, M. J. (1983) *Housing Policy and Economic Power*, London, Methuen.

—— (1985) 'The urban rent question', *Environment and Planning A* 17 (4): 503–25.

Brindley, T., Rydin, Y. and Stoker, G. (1989) *Remaking Planning*, London, Unwin Hyman.

Comedia (1991) *Out of Hours*, London, Gulbenkian Foundation.

Commission of the European Communities (1990) *Green Paper on the Urban Environment*, COM(90), 218 final, Brussels, European Commission.

—— (1991) *Europe 2000*, Brussels, European Commission.

CBI (Confederation of British Industry) (1991) *A London Development Agency: Optimising the Capital's Assets*, London, CBI.

Coopers & Lybrand Deloitte (1991) *London World City: Stage II Report*, London, Coopers & Lybrand Deloitte.

DoE (Department of the Environment) (1988) *Creating Development Trusts*, London, HMSO.

—— (1989) *Strategic Guidance for London*, London, DoE.

GLC (1984) *Draft Greater London Development Plan*, London, GLC.

Hall, P. (1989) *London 2001*, London, Unwin Hyman.

—— (1991) 'A new strategy for the South East', *The Planner* 77(10), 22 March: 6–9.

Harvey, D. (1982) *The Limits to Capital*, Oxford, Blackwell.

—— (1985) *The Urbanization of Capital*, Oxford, Blackwell.

Hillman, J. (1988) *A New Look for London*, London, HMSO.

King, A. D. (1990) *Global Cities*, London, Routledge.

Labour Party (1990) *Looking to the Future*, London, Labour Party.

—— (1991) *London: A World Class Capital*, London, Labour Party.

Leitner, M. (1990) 'Cities in pursuit of economic growth: the local state as entrepreneur', *Political Geography Quarterly* 9 (2): 146–80.

Lloyd, G. and Newlands, D. (1990) 'Business interests and planning initiatives: a case study of Aberdeen', in J. Montgomery and A. Thornley (eds) *Radical Planning Initiatives*, Aldershot, Gower.

London Planning Advisory Committee (1988) *Strategic Planning Advice for London*, London, LPAC.

Luithlen, L. (1986) 'Interest-bearing capital and commercial property investment', in *The Production of the Built Environment*, Proceedings of the 7th Bartlett International Summer School, London, University College London.

MacDonald, K. and Newman, P. (1991) 'Paris planners sketch in a new regional structure', *Planning* 916:18–19.

Montgomery, J. (1990) 'Cities and the art of cultural planning', *Planning Practice & Research* 5 (3): 17–24.

Nabarro, R. (1990) 'The investment market in commercial and industrial development', in P. Healey and R. Nabarro (eds) *Land and Property Development in a Changing Context*, Aldershot, Gower.

Robinson, S. (1990) 'City under stress', *The Planner* 76(10): 10.

Ryser, J. (1991) *Germany in Transition: New Strategies of Urban Development: Berlin/London*, London, Goethe Institute.

Savitch, H. V. (1988) *Post-Industrial Cities*, Princeton, Princeton University Press.

Tuckett, I. (1990) 'Coin Street: there is another way', in J. Montgomery and A. Thornley (eds) *Radical Planning Initiatives*, Aldershot, Gower.

INDEX